HOUSING, FUEL POVE

We who are liberal and progressive know that the poor are our equals in every sense except that of being equal to us.
Lionel Trilling, 'The Liberal Imagination' (1950)

Housing, Fuel Poverty and Health
A Pan-European Analysis

JONATHAN D. HEALY
Combat Poverty Agency, Dublin

LONDON AND NEW YORK

First published 2004 by Ashgate Publishing

2 Park Square, Milton Park, Abingdon, Oxon OX14 4RN
711 Third Avenue, New York, NY 10017, USA

Routledge is an imprint of the Taylor & Francis Group, an informa business

First issued in paperback 2016

Copyright © 2004 Jonathan D. Healy

Jonathan D. Healy has asserted his right under the Copyright, Designs and Patents Act, 1988, to be identified as the author of this work.

All rights reserved. No part of this book may be reprinted or reproduced or utilised in any form or by any electronic, mechanical, or other means, now known or hereafter invented, including photocopying and recording, or in any information storage or retrieval system, without permission in writing from the publishers.

Notice:
Product or corporate names may be trademarks or registered trademarks, and are used only for identification and explanation without intent to infringe.

British Library Cataloguing in Publication Data
Healy, Jonathan D.
   Housing, fuel poverty and health : a pan-European analysis
   1.Poverty - Europe 2.Poverty - Ireland 3.Poverty - Health
   aspects - Europe 4.Poverty - Health aspects - Ireland
   5.Dwellings - Energy consumption - Europe 6.Dwellings -
   Energy consumption - Ireland 7.Power resources - Europe -
   Costs 8.Power resources - Ireland - Costs 9.Europe - Social
   conditions - 20th century 10.Ireland - Social conditions -
   20th century
   I. Title
   362.5'094

Library of Congress Cataloging-in-Publication Data
Healy, Jonathan D.
   Housing, fuel poverty and health : a pan-European analysis / by Jonathan D. Healy.
      p. ; cm.
   Includes bibliographical references and index.
   ISBN 0-7546-4218-6
   1. Poverty--Health aspects--Europe. 2. Health status indicators--Europe. 3. Poor--Energy assistance--Europe. 4. Poor--Housing--Europe. 5. Poverty--Health aspects--Ireland. 6. Health status indicators--Ireland. 7. Poor--Energy assistance--Ireland. 8. Poor--Housing--Ireland.
   [DNLM: 1. Health Status--Europe. 2. Fossil Fuels--economics--Europe. 3. Health Policy--Europe. 4. Housing--economics--Europe. 5. Poverty--Europe. WB 141.4 H4347h 2004] I. Title.

   RA418.5.P6H44 2004
   614.4'24--dc22
                                                                                    2004013250

ISBN 978-0-7546-4218-3 (hbk)
ISBN 978-1-138-26662-9 (pbk)

Transfered to Digital Printing in 2013

Contents

List of Figures	vi
List of Tables	viii
Glossary	xii
Acknowledgements	xvi
1. Introduction	1
2. Housing Deprivation in the EU	7
3. Fuel Poverty in the EU: Developing a Composite Measurement	32
4. The Severity of Fuel Poverty in Ireland	65
5. Housing Deprivation and Self-Reported Health in the EU	87
6. Fuel Poverty and Health in Ireland	106
7. Fuel-Poor Households and Risk Factors	119
8. Fuel Poverty, Thermal Comfort and Household Occupancy in Ireland	129
9. Excess Winter Mortality in the EU: Identifying Key Risk Factors	140
10. Policy Implications, Strategies and Recommendations	161
11. Summary and Conclusions	191
Appendix I: Regression Results	208
Appendix II: Regression Results	232
Bibliography	235
Index	249

List of Figures

2.1	Energy consumption per household ('000 MJ, 1996)	22
2.2	Housing conditions: cumulative results (mean %, 1994-97)	30
4.1	Reasons for non-investment in energy-saving measures in Ireland	82
4.2	Fuel poverty in Ireland: 10 risk groups	86
5.1	Poor health and ability to heat home adequately (mean % of households, 1994-97)	91
5.2	Poor health and adequate heating facilities (mean % of households, 1994-97)	93
5.3	Poor health and damp (mean % of households, 1994-97)	94
5.4	Poor health and rot (mean % of households, 1994-97)	95
5.5	Poor health and central heating (mean % of households, 1994-97)	96
5.6	Poor health and leaky roofs (mean % of households, 1994-97)	97
5.7	Poor health and overcrowding (mean % of households, 1994-97)	98
5.8	Poor health and burdensome housing costs (mean % of households, 1994-97)	100
5.9	Poor health and late payment of utility bills (mean % of households, 1994-97)	101
5.10	Poor health and housing satisfaction (mean % in each group, 1994-97)	102
9.1	Seasonal variation in mortality (SVM) and mean winter temperature in EU-14	146
9.2	Seasonal variation in mortality (SVM) and PPP-adjusted GDP per capita in EU-14	148
9.3	Seasonal variation in mortality (SVM) and health expenditure as a % of GDP per capita in EU-13	149
9.4	Seasonal variation in mortality (SVM) and public health expenditure as a % of GDP per capita in EU-13	150
9.5	Seasonal variation in mortality (SVM) and public health expenditure (PPP-adjusted $) in EU-13	151
9.6	Seasonal variation in mortality (SVM) and hospital beds (per 1,000 population) in EU-13	151
9.7	Seasonal variation in mortality (SVM) and income poverty in EU-12	153
9.8	Seasonal variation in mortality (SVM) and income inequality in EU-12	154
9.9	Seasonal variation in mortality (SVM) and multiple deprivation in EU-12	154
9.10	Seasonal variation in mortality (SVM) and fuel poverty in EU-14	155
9.11	Seasonal variation in mortality (SVM) and penetration of cavity-wall insulation in EU-13	157

List of Figures

9.12 Seasonal variation in mortality (SVM) and penetration of double-glazing in EU-13	157
11.1 Composite fuel poverty in EU-14 (% of households, 1994-97)	194
11.2 Ability to heat the home adequately by severity (% of households, Ireland, 2001)	195
11.3 Chronic fuel poverty in Ireland, 1994-2001 (% of households)	196
11.4 Fuel poverty in Ireland: 10 risk groups	197
11.5 Living-room temperatures of fuel-poor, elderly and all other households (% households, Ireland, 2001)	202
11.6 Relative excess winter mortality in EU-14 (mean, 1988-97)	204

List of Tables

1.1	Country acronyms used in this study	2
2.1	Availability of panel data in EU-15	9
2.2	Population levels (millions) (P), number of dwellings (millions) (D) and mean occupancy (persons per household) (O), 1996	10
2.3	Dwelling types in Europe (mean, 1994-97): % and number of households (millions)	11
2.4	Housing tenure in Europe (mean, 1994-97): % and number of households (millions)	13
2.5	Occupancy levels (mean persons per household, 1994-97): % and number of households (millions)	14
2.6	Households with leaking roofs (1994-97): % and number of households affected ('000)	15
2.7	Households lacking hot running water (1994-97): % and number of households affected ('000)	16
2.8	Households objectively overcrowded (1994-97): % and number of households affected ('000)	17
2.9	Households subjectively overcrowded (1994-97): % and number of households affected ('000)	18
2.10	Damp households (1994-97): % and number of households affected ('000)	19
2.11	Households with rotten window frames (1994-97): % and number of households affected ('000)	20
2.12	Percent of households with various energy-efficiency measures (1996)	21
2.13	Households whose housing costs are a "heavy" financial burden (1994-97): % and number of households affected ('000)	24
2.14	Households unable to heat home adequately (1994-97): % and number of households affected ('000)	25
2.15	Households unable to pay scheduled utility bills (1994-97): % and number of households affected ('000)	26
2.16	Tenant households that pay for home heating in addition to rent (1994-97): % and number of households affected ('000)	27
2.17	Households "totally" or "very dissatisfied" with housing (1994-97): % and number of households affected ('000)	28
3.1	Sample size and response rate for ECHP in 1994	37
3.2	Households unable to heat home adequately (% of households)	40
3.3	Households unable to pay scheduled utility bills (% of households)	41
3.4	Households declaring a lack of adequate heating facilities (% of households)	42

List of Tables ix

3.5	Households with damp spores (% of households)	44
3.6	Households without central or electric-storage heating (% of households)	45
3.7	Households with rotten window frames (% of households)	46
3.8	Sensitivity analysis of fuel poverty results (% of households, 1994-97)	49
3.9	Households unable to heat home adequately by socio-demographic group (mean % of households, 1994-97)	52
3.10	Fuel poverty in Europe by housing type (mean % of households, 1994-97)	53
3.11	Fuel poverty in Europe by marital status (mean % of households, 1994-97)	55
3.12	Fuel poverty by educational attainment (mean % of households, 1994-97)	56
3.13	Fuel poverty in Europe by main income source (mean % of households, 1994-97)	57
3.14	Fuel poverty in Europe by housing tenure (mean % of households, 1994-97)	58
3.15	UK fuel poverty in 1996: a comparison using alternative approaches	59
4.1	Percentage of Irish households with various energy-saving measures (1996-2001)	66
4.2	Ability to adequately heat home: incidence and numbers affected (Ireland, 2001)	68
4.3	Ability to heat the home adequately by marital status (% households)	69
4.4	Ability to heat the home adequately by marital status of parents (% households)	70
4.5	Ability to heat the home adequately by region (%)	70
4.6	Ability to heat the home adequately by educational attainment (% households)	71
4.7	Ability to heat the home adequately by dwelling age (% households)	72
4.8	Ability to heat the home adequately by housing occupancy (number of persons)	72
4.9	Ability to heat the home adequately by social class (% households)	74
4.10	Ability to heat the home adequately by main income source (%households)	74
4.11	Ability to heat the home adequately by employment status (%households)	75
4.12	Ability to heat the home adequately by whether householder ever had paid work (% households)	76
4.13	Ability to heat the home adequately by income (% households)	76
4.14	Ability to heat the home adequately by number of dependent children (% households)	77
4.15	Ability to heat the home adequately by housing tenure (% households)	78
4.16	Ability to heat the home adequately and households claiming fuel subsidies (% households)	79
4.17	Ability to heat the home adequately and presence of damp (%households)	80

4.18	Ability to heat the home adequately and presence of condensation (% households)	80
4.19	Predictive power of the Probit regression for ability to heat home adequately (%)	81
5.1	Summary of statistical analysis on deprivation indicators and poor Health	104
5.2	Correlation coefficients and significance levels for deprivation levels and poor health	104
6.1	Fuel poverty and self-reported health status	108
6.2	Fuel poverty and self-reported health status now compared to 1 year ago	108
6.3	Fuel poverty and manifestations of poor physical health	109
6.4	Fuel poverty and manifestations of poor emotional health	110
6.5	Fuel poverty and feelings	110
6.6	Fuel poverty and self-perceived health status	111
6.7	Fuel Poverty and objective health outcomes	112
6.8	Fuel poverty and related chronic health outcomes	114
6.9	Fuel poverty and worries	115
6.10	Fuel poverty and quality of life/life satisfaction	116
7.1	Fuel poverty and winter time spent outdoors	121
7.2	Fuel poverty and outdoor cold exposure	122
7.3	Fuel poverty and duration of outdoor cold exposure	122
7.4	Fuel poverty and duration of indoor cold exposure	123
7.5	Fuel poverty and lifestyle risk factors	124
7.6	Fuel poverty and diet	124
7.7	Fuel poverty and levels of physical activity	125
7.8	Fuel poverty and usual mode of transport	126
8.1	Fuel poverty and % of households with sub-optimal thermal comfort	133
8.2	Fuel poverty and living room temperature	135
8.3	Fuel poverty and duration of indoor cold strain	136
8.4	Fuel poverty and who occupies house most	137
8.5	Fuel poverty and manifestations on occupancy	138
9.1	Weather stations employed for climatic data	144
9.2	Relative excess winter mortality in EU-14 (1988-97)	145
9.3	Relative excess winter mortality and climatic factors in EU-14	147
9.4	Relative excess winter mortality and macroeconomic variables in EU-14	148
9.5	Relative excess winter mortality and lifestyle risk factors in EU-14	152
9.6	Relative excess winter mortality and health indicators in EU-12	153
9.7	Relative excess winter mortality and socio-economic variables in EU-14	155
9.8	Relative excess winter mortality and domestic thermal efficiency in EU-13	156
9.9	Results of Poisson regression model: cross-country relationships with relative excess winter mortality (regression coefficients and significance)	159

10.1	Total exchequer expenditure on the 'fuel allowance' in Ireland, 1992-2002	174
10.2	Summary of key findings and recommended policy strategies	190

Appendix I

1.	Ability to heat home adequately: results of a Probit model for 1994 and 1995	208
2.	Ability to heat home adequately: results of a Probit model for 1996 and 1997	210
3.	Ability to pay utility bills in past year: results of a Probit model for 1994 and 1995	212
4.	Ability to pay utility bills in the past year: results of a Probit model for 1996 and 1997	214
5.	Households lacking adequate heating facilities: results of a Probit model for 1994 and 1995	216
6.	Households lacking adequate heating facilities: results of a Probit model for 1996 and 1997	218
7.	Presence of central heating: results of a Probit model for 1994 and 1995	220
8.	Presence of central heating: results of a Probit model for 1996 and 1997	222
9.	Presence of damp: results of a Probit model for 1994 and 1995	224
10.	Presence of damp: results of a Probit model for 1996 and 1997	226
11.	Presence of rot: results of a Probit model for 1994 and 1995	228
12.	Presence of rot: results of a Probit model for 1996 and 1997	230

Appendix II

1.	Ability to heat home to an adequate level: results of an ordered-Probit model	232

Glossary

adverse selection	The phenomenon that results from the inability of one trader to assess the quality of another (due to information asymmetries) which makes it likely that poor-quality traders will predominate.
Coase Theorem	A theorem often used in resource and environmental economics that states economic efficiency is achieved when property rights are fully allocated and when free trade of all property rights is possible.
'consensual' approach	The Consensual approach to measuring poverty employs indicators of socially perceived necessities. Unlike traditional forms of measuring relative poverty, this approach does not rely on the opinions or scientific postulates of academics or experts.
Cost-Benefit Analysis	The appraisal of an investment project which includes all private and social costs accruing to the project over a pre-identified time period.
DEFRA	Department of Environment, Food and Rural Affairs (UK)
DTI	Department of Trade and Industry (UK)
endogenous variables	A variable whose values are determined by other variables within a system.
energy paradox	The slow, gradual diffusion across the housing stock of economically sound energy-saving measures.
EU-14	All Member States of the European Union during the years 1994 to 1997, including: Austria, Belgium, Denmark, Finland, France, Germany, Greece, Ireland, Italy, Luxembourg, the Netherlands, Portugal, Spain and the United Kingdom.

Glossary xiii

ex ante	Expected or intended beforehand. The terms is often used in economic evaluation to denote project appraisal prior to the implementation of the project.
ex post	The result afterwards. The term is often used in economic evaluation to denote project appraisal after the project has been implemented.
excess winter mortality	The increased number of deaths which occur during the winter months over and above the mean mortality rate of the non-winter seasons.
exogenous variables	Variables whose values are not determined within the set of equations or models established to make predictions or test a hypothesis.
externalities (external costs/benefits)	The non-market costs (or benefits) of an action borne not by the private individual instigating the action but by society at large.
free-rider problem	The problem arising from an instance when no individual is willing to contribute towards the cost of a public good when it is hoped that someone else will bear the cost instead.
fuel poverty	The inability to afford adequate home heating, or, the inability to heat the home adequately for 10 per cent or less of household income.
Home Energy Efficiency Scheme (HEES)	The UK Government's chief policy response to the alleviation of fuel poverty over the past decade which includes means-tested grants to low-income and elderly households to improve the energy-efficiency characteristics of their homes.
information asymmetry	A situation in which typically buyers and sellers possess differing levels of information about a good or service which generally results in market failure.

longitudinal analysis	An empirical analysis of trends in which dynamic effects (over time) are highlighted.
market failure	An outcome deriving from the self-interested behaviour of individuals in the context of free trade in which economic efficiency does not result and which provides ubiquitous argument for intervention of some kind.
NEA	Neighbourhood Energy Action
Net Social Benefit (NSB)	The monetary benefit (if positive) accruing to society as a result of a proposed project investment usually obtained through economic evaluation techniques such as Cost-Benefit Analysis.
OPEC	Organisation for Petroleum Exporting Countries
payback period	The period over which the cumulative net revenue from an investment project equals the original investment.
poverty trap	The combination of losing state benefit (social welfare) and paying income tax ensures that poor families keep very little of any extra money they earn.
Probit	This regression model extends the principles of generalised linear models to treat the case of dichotomous and polytomous dependent variables. These methods differ from standard regression in substituting maximum likelihood estimation of a link function of the dependent for regression's use of least squares estimation of the dependent itself. The function used in Probit is the inverse of the standard normal cumulative distribution function.
regression analysis	A mathematical and statistical (econometric) technique commonly employed in applied social-science research for estimating the parameters of an equation from sets of data of the independent and dependent variables.

subsidy	A government grant to suppliers of goods and services.
tradeable permit	A quantity-based economic instrument that facilitates compliance with environmental emissions' quotas (such as those set at the Kyoto and Gothenburg Protocols) through the buying and selling of emissions' permits in a competitive marketplace.
transactions' costs	The costs associated with the process of buying and selling, sometimes referred to as 'hassle costs'.

Acknowledgements

I want to express my sincere thanks to Frank Convery for his enthusiasm, passion and support for this research, and also for his thorough reading of an early draft. Frank is a truly inspiring person and I am very indebted to him for his encouragement of my efforts in his Department at both master's and doctoral levels. I'd like also to thank Peter Clinch for his efforts (during hard times) in obtaining funding for the purchase of the datasets employed in this study, his work on the Probit analysis and the national household survey, and his comments on the first draft. Thanks to Paul Ekins for subsequent detailed comments and to Orla Lane for helpful notes and corrections on the policy chapter. I would like to thank several anonymous referees for very useful suggestions and remarks which have benefited the final version of this book. For their efficiency and courtesy I wish to thank the Ashgate Publishing team, especially Alison Kirk. I also want to thank my former UCD colleague Martin Sokol for his support, advice and much-needed good humour during my completion of the manuscript, and thanks also to my current colleagues in the CPA, most notably Jim Walsh for some insights on the issue of carbon taxation which I borrowed for use in the final draft! I am also very grateful to those who supported this research financially, including the Irish Research Council for the Humanities and Social Sciences, the Environmental Institute and Urban Institute Ireland, University College Dublin. Finally, a big Cheers! to all my family and friends for being a mountain of support.

Chapter 1

Introduction

Why This Book?

The principal aim of this book is to examine the relationship between domestic energy efficiency, fuel poverty and related health impacts using a comparative study framework. This study tests a number of principal hypotheses using European data. First, fuel poverty and poor domestic energy inefficiency is examined to assess its relationship with impaired health status. The second hypothesis asserts that fuel poverty and poor domestic energy inefficiency is strongly associated with high levels of excess winter mortality. The third hypothesis argues that a 'Consensual' approach to calculating fuel poverty results in more conservative, but more reliable, estimates than the standard definition. The fourth hypothesis examines whether the fuel-poor exhibit higher risk factors associated with poor health than others. The book then turns to issues of thermal comfort and examines whether the fuel-poor endure reduced thermal comfort in the home and lower ambient household temperatures. The final key hypothesis concerns the relationship between fuel poverty and excess winter deaths. In the final data analysis chapter, seasonal variations in mortality are examined against a wide range of social and economic factors, not just fuel poverty and inadequately insulated homes, to identify causality.

The book also pays particular attention to the case of Ireland, a country known to suffer from high levels of fuel poverty. Some additional hypotheses are tested in this respect. First, a hypothesis is tested that domestic thermal- (energy-) efficiency standards in Ireland are among the poorest in Europe. The book also examines whether Ireland exhibits the poorest housing conditions and the highest levels of housing deprivation, unaffordable housing and dissatisfaction with housing in the EU. A third hypothesis relates to whether fuel poverty in Ireland is among the highest in Europe. The study also focuses on health impacts by testing formally the notion that health impairment associated with fuel poverty in Ireland is among the highest Europe; the Irish level of excess winter mortality is also regressed formally to identify key risk factors. Finally, it is hypothesised that the information gap is the main reason for market failure in Irish domestic energy efficiency.

The study quantifies levels of fuel poverty and seasonal variations in mortality in 14 EU Member States and investigates empirically the relationship between fuel poverty and health status.[1] The Member States are:

Table 1.1 Country acronyms used in this study

Country	Acronym
Germany	D
Denmark	DK
Netherlands	NL
Belgium	B
Luxembourg	L
France	F
United Kingdom	UK
Ireland	IRL
Italy	I
Greece	EL
Spain	E
Portugal	P
Austria	A
Finland	FIN

The work is made up of two major methodological components: an analysis of a national household survey of Ireland conducted in 2001, and an analysis of longitudinal datasets from the European Community Household Panel (ECHP) and other data obtained chiefly through the United Nations and the World Bank. More precisely, the main components of this book are as listed below.

The book begins in Chapter 2 with a comparative assessment of housing conditions, thermal-efficiency standards, housing deprivation, housing affordability and housing dissatisfaction in EU-14.[2] Cross-country levels of fuel poverty in EU-14 are quantified using a new composite measure based on longitudinal social indicators in what can be termed a 'Consensual' approach to calculating fuel poverty; a full chapter is dedicated to describing this approach and the results. The health impacts of fuel poverty in EU-14 are then examined empirically in a number of chapters. Levels of income poverty and income inequality in EU-14, as well as incidences of multiple deprivation, are then quantified. Seasonal variations in mortality in EU-14 are calculated. The risk factors associated with seasonal variations in mortality across EU-14 are then identified formally using Poisson regression analysis.

---

[1] Data on the 15th Member State, Sweden, are only available for certain variables in 1997.
[2] Note again that Sweden is excluded from most of the analysis due to its late adoption of the ECHP.

Introduction   3

The study also examines empirically the main reasons for non-adoption of energy-saving measures using Ireland as a case study. As mentioned, a large section of the book examines empirically, and in detail, the health outcomes associated with fuel poverty. In this regard, an investigation is undertaken in order to determine whether fuel-poor households demonstrate higher risk factors than other households. Furthermore, the current ambient temperatures of housing are measured using an Irish sample. Thermal comfort among the fuel-poor and others is compared using new empirical data. In addition, adverse occupancy issues relating to fuel poverty are also investigated.

The study is highly policy relevant and, for this reason, a large policy chapter is included in the book which outlines a number of major policy implications of the study. International policy experience in this field is examined, and a number of *ex post* exemplars are highlighted. The key economic (and other) instruments available to policymakers to internalise a number of externalities is then outlined briefly. Policy recommendations are proffered at the chapter's close.

Context

Ireland is an interesting case study within the context of EU-14 for a number of reasons. The housing stock appears to be one of the most energy inefficient in northern Europe,[3] as Chapter 4 of this study illustrates.[4] In addition, Ireland is thought to demonstrate among the highest rates of fuel poverty and excess winter mortality in northern Europe – both of which are hypothesised to be strongly related to energy inefficient housing (Clinch and Healy, 2000c; Eurowinter Group, 1997). Because of this energy inefficiency, energy consumption in the domestic sector is greater than necessary, as people inhabiting inefficient dwellings must consume more energy to heat their homes.[5] Consequently, environmental emissions are also greater. This is of considerable importance given that Ireland is having extreme difficulty in meeting its agreed target for stabilisation of greenhouse-gas emissions under the Kyoto Protocol and acidification precursors under the Gothenburg Agreement.

Fuel poverty, formally first identified in the early 1970s, has continued to exist as a significant social ill three decades on. The debate on fuel poverty in the

---

[3] 'Northern Europe' is defined broadly as the ten non-Mediterranean countries in the study (i.e. excluding 'Southern'/Mediterranean nations including Italy, Greece, Spain and Portugal).

[4] 'Energy inefficient' in this case is defined in the housing-engineering (technical) sense as dwellings lacking energy-saving measures such as double-glazing, cavity-wall insulation, draught stripping and so forth. However, this study uses differing definitions of the term depending on the context in which it is employed. For instance, an economic definition of 'energy efficient' might be "that level of domestic energy efficiency at which the net benefit to society is maximised", whilst a purely financial definition would specify the level of energy efficiency that maximises the net financial gain to the individual householder.

[5] It has also been shown that the poorest individuals tend to spend three times more than the average on energy relative to income (Clinch and Healy, 1999a).

1980s and 1990s, where it had been argued that affordable warmth is a "right rather than a privilege", has become relevant once again in the light of unstable and rising worldwide fuel prices. Fuel poverty is different to poverty. Poverty can be eradicated through income support (as has been noted by Boardman, 1991), whereas the eradication of fuel poverty requires not just income subsidisation but also crucial investment in the capital stock (i.e. the household), as fuel poverty is caused by a complex interaction between low income and domestic energy inefficiency.

When fuel becomes more expensive (as it has done intermittently in Europe in particular over the past two to three years),[6] some households find it increasingly difficult to heat their homes adequately. Those living in energy inefficient dwellings and on low incomes often cannot afford to heat their homes adequately and suffer from fuel poverty. The problem is a serious political, environmental, social and public-health issue and the UK government, in particular, has invested considerably in assisting low-income groups upgrade their housing. However, recent research confirmed the persisting nature and considerable scale of the problem in the UK (DEFRA and DTI, 2001). In addition, fuel poverty has many attendant effects, most notably on human health (Clinch and Healy, 2000c; Rudge and Nicol, 2000). Yet, the net benefits to society (in terms of reduced energy consumption and environmental emissions and improvements in health and comfort) of eradicating fuel poverty through implementing domestic energy-efficiency programmes would be very substantial (Blasnik, 1998; Clinch and Healy, 2001).

The UK government is now obliged by statute to eliminate fuel poverty by 2016. To this end, a detailed fuel poverty strategy has been published which describes how it intends to meet this ambitious goal (DEFRA and DTI, 2001). However, responsibility for the key policy tools needed for meeting the target rests with different levels of the UK administration. Some are decided at the UK level (e.g. benefit levels), some at the British level (e.g. energy policy), and some at the devolved administration level (the Scottish Parliament, for instance) (e.g. energy efficiency policies). Furthermore, responsibility for collecting the data necessary for establishing whether the Government is going to meet its target lies with the devolved administrations. Unfortunately, data collection standards between the different administrations vary widely.

If the UK Government is having problems arriving at a common standard for assessing the level of fuel poverty in the UK, then the difficulty in reaching a common understanding across Europe, particularly when the term has yet to be accepted in most European countries, is unambiguous. The UK Government appears very keen for the issue to be more widely acknowledged within Europe, not least because it raises the possibility of attracting major European funds towards tackling the problem. It is therefore looking for evidence that might show the phenomenon is indeed widespread across Europe. This work is intended to improve the knowledge base in this regard at an EU-wide level.

---

[6] Although some countries, such as the UK, have experienced real declines in energy prices.

Furthermore, there has been virtually no comparative empirical analysis of fuel poverty undertaken heretofore. This research deficit has occurred, not because of a lack of interest in the area, but rather as a result of some logistical factors, most obviously the lack of comparable cross-country data.

With these issues in mind, a study was undertaken to evaluate a programme to retrofit the entire housing stock in Ireland with insulation and heating measures so as to bring it to the thermal standards of the latest building regulations.[7] The study contained the most comprehensive economic analysis of domestic energy-conservation opportunities in Ireland and found that the retrofitting programme would result in a net benefit to society of some €3.1 billion with a benefit-cost ratio of 3:1, a payback period of just 7 years and an internal rate of return of 33%. Furthermore, this analysis found that the potential health benefits of improving the housing stock (reduced excess winter mortality and morbidity) accounted for the second-largest programme benefit (after energy savings) at €1.2 billion. As such, this was the first in-depth analysis of housing and health in Ireland. However, there were a number of caveats and limitations with the study's health analysis – many owing to temporal and financial constraints – and, consequently, some of the results of the analysis concerning impacts on human health have to be described and treated cautiously.

A recent review of the evidence on health effects of improving housing has found that, despite many studies showing some health gains after an intervention, "the small populations and lack of controlling for confounding factors limit the generalisability of these findings" (Thomson *et al.*, 2001). The authors go on to point out that the lack of evidence linking housing and health may be attributable to pragmatic difficulties with housing studies as well as the political climate in the UK. They conclude that,

> a holistic approach is needed that recognises the multi-factorial and complex nature of poor housing and deprivation. Large-scale studies that investigate the wider social context of housing interventions are required.

The *raison d'être* of this study is to probe the association between cold housing and health by developing a more holistic approach that attempts to control for a number of confounding factors by incorporating a large number of social and economic risk factors into a large-scale, cross-country analysis.

---

[7] The economic analysis for the study conducted by this author formed the bulk of his M.Sc. thesis. An overview of the economic results can be found in Clinch and Healy (2001), while readers should consult Brophy *et al.* (1999) for the full commissioned report.

Outline of Book

The results of the study regarding housing conditions, deprivation, affordability and general satisfaction with housing are presented comparatively in Chapter 2. Chapter 3 describes the cross-country results for fuel poverty using the new composite measurement developed in this study. A socio-economic analysis is presented in this chapter, and Probit multivariate regression models test the strength of significance of the relationships found. The national household survey is employed in Chapter 4 to quantify the severity of fuel poverty in Ireland, its relationship with adverse housing conditions and reasons for non-investment in energy-saving measures. Chapter 5 presents the results of the study focusing on the health impacts associated with fuel poverty in EU-14. These are dealt with in more detail in Chapter 6 where the national household survey is utilised to quantify the specific adverse physical and emotional health outcomes of fuel poverty and domestic energy inefficiency. Chapter 7 investigates empirically whether Irish fuel-poor households demonstrate higher levels of risk factors associated with impaired health. A profile of the thermal comfort of housing and the relationship between fuel poverty and ambient temperature and thermal comfort is discussed in Chapter 8. Attendant adverse occupancy issues associated with fuel poverty are also analysed using the survey data. This chapter places particular focus on the impacts of fuel poverty on older people. The final results chapter (Chapter 9) presents the results of the model testing for risk factors associated with variations in seasonal mortality. Relative excess winter mortality is calculated across EU-14 and key risk factors are identified. The main findings of the study are summarised in Chapter 10. Chapter 11 presents the policy implications of the study and summarises policy interventions on fuel poverty in other countries before presenting country-by-country policy recommendations. An Appendix is included in the book containing the questions used in the national household survey and European Community Household Panel.

While the book has been edited so that it may be read from start to finish as one study, Chapters 2 to 9 inclusive have been specifically designed to be read as stand-alone documents. This style of presentation enables the reader to read any given chapter as a stand-alone piece of research. The methodology and existing literature are discussed succinctly in each chapter. However, repeated material is kept to an absolute minimum, enabling those wishing to read the book from beginning to end to do so without re-reading unnecessary material.

Chapter 2

Housing Deprivation in the EU

Introduction

Housing standards, especially those pertaining to energy efficiency, vary considerably across Europe, as has been shown in a provisional analysis by Healy (2001a). Of course, certain countries prioritise thermal measures in the design and construction of new housing, as it is essential protection to combat the relatively severe winters experienced in these colder climates where winter temperatures are often below freezing (Clinch and Healy, 1999a). Ireland and the UK have the highest rates of seasonal mortality in northern Europe, and it has been shown that such mortality rates result, in no small part, from the inadequately protected, thermally inefficient housing stocks in these countries (Curwen, 1991; Eurowinter Group, 1997). There are also strong associations between inadequately heated homes and increased rates of morbidity; higher incidences of various cardiovascular and respiratory diseases have been associated with cold exposure from within the home (Collins, 1986; Evans *et al.*, 2000). Thus, when temperatures fall during a typical British or Irish winter, households need to increase their expenditure on fuel considerably to heat their home adequately, owing to the poor level of heat retention in their dwellings. The problem of fuel poverty occurs, therefore, when a household does not have the adequate financial resources to meet these winter home-heating costs, and/or because the dwelling's heating system and insulation levels prove to be inadequate for achieving affordable household warmth. The link between poor housing standards, fuel poverty and ill health is a matter that is currently being given much (needed) research.[1]

Data on housing conditions and energy-efficiency levels in southern Europe are notoriously difficult to obtain, and their reliability is often questionable. However, a provisional analysis of general housing conditions, conducted by Healy (2001a), indicated that high levels of energy efficiency in southern-European housing are not prioritised in building regulations. It is an often-overlooked fact that many southern-European countries also face cold winter temperatures, yet their housing stocks appear to be poorly protected from the cold, and they are also the poorest countries in the European Union.[2,3] Despite this, there has been

---

[1] Rudge and Nicol (2000) present an overview of current research in this area in the UK, while Chapters 7 and 8 of this thesis present the results for Europe.
[2] Calculated using measures such as income poverty and inequality as well as macro-economic indicators like GDP per capita
[3] Italy being the exception here, with GDP per capita considerably above EU-average.

virtually no published research on housing conditions and fuel poverty in southern Europe, although there is a growing body of research on this area in the UK and Ireland. This analysis attempts to rectify this research deficit.

From an environmental-policy perspective, many countries demonstrating poor levels of domestic energy efficiency are consuming greater amounts of energy than necessary, as individuals inhabiting inefficient dwellings must consume more fuel to heat their homes. This is of considerable importance given that many European countries are having extreme difficulty in meeting their agreed targets for stabilisation of greenhouse-gas emissions under the Kyoto Protocol and acidification precursors under the Gothenburg Agreement (Clinch, 2001).

In light of these policy-drivers, this chapter presents a cross-country analysis of housing in 14 European countries. It examines a variety of housing conditions experienced in each of these countries, energy-efficiency levels, householder finances and general satisfaction with housing. This is done using new, longitudinal datasets from the European Statistical Office's European Community Household Panel (ECHP). The data cover the four years, 1994-1997 respectively.

The first group of results provides general contextual cross-country data on dwelling specifications (type), housing tenure and household size (occupancy levels). There then follows a detailed discussion on housing conditions in which the proportion of households demonstrating the lack of essential amenities (e.g. hot running water) and the presence of unwanted housing conditions (e.g. leaky roofs, overcrowding, damp, etc.) is isolated on a country-by-country basis across time; absolute numbers of households affected are also quantified. Data on domestic energy-efficiency levels and household energy consumption are illustrated using a complementary Eurostat survey.[4]

There is an increasing literature analysing housing affordability as a key indicator of housing conditions and householder happiness.[5] Moreover, research on quality-of-life indicators has often stressed the importance of housing satisfaction on happiness, especially among the elderly (Barresi *et al.*, 1984; Kozma and Stones, 1983). In light of this, the analysis of housing conditions in this chapter also extends to household finances and satisfaction. Householders are asked to declare how burdensome their housing costs are on a scale of one to five, whether they can afford adequate home heating (or are, otherwise, 'fuel-poor') and whether they can pay their utility bills on time. Furthermore, rent-paying tenant householders are asked whether heating costs are paid in addition to rent; this latter result may be considered useful in assessing the levels of fuel subsidisation and degree of hardship facing fuel-poor households. Finally, levels of housing dissatisfaction are reported. A cumulative result (based on ten key indicators of housing conditions) is presented for each country which shows a strong pattern of deprivation. Some key policy implications are drawn at the end of the chapter.

---

[4] See Eurostat (1999). Data on energy-efficiency levels and energy consumption are not provided in the ECHP.
[5] See, for example, studies by Konadu-Agyemang (2001) and Rakodi and Withers (1995).

European Community Household Panel (ECHP)

The ECHP is a standardised, multi-purpose and longitudinal survey, providing comparable information across European-Union Member States on income, work and employment, poverty and social exclusion, housing, health and other diverse social indicators regarding the living conditions of private households and persons. The crucial feature of the ECHP is the harmonisation of its methodology and results, through the creation of a centralised questionnaire. During the first 'wave' of the questionnaire, the collection of data occurred in 12 countries in Europe (all EU Member States in 1994); this increased to 13 in 1995 (when Austria joined), 14 in 1996 (with the inclusion of Finland), and 15 in 1997 (Sweden is now on board). Consequently, some flexibility was granted to each participating country to adapt common procedures to acquiesce with their own local situations.

Data are collected by National Data Collection Units in each country. These Units – normally research institutes or national statistics' centres – tailor the questionnaire to make it suitable for their own respective countries. High response rates of about 70% were obtained for the first four years of the survey, and some 60,000 households and approximately 130,000 adults are interviewed successfully in each wave. The data used in this chapter come from the first four years of the ECHP (i.e. 1994-1997). More details of the survey can be found in Eurostat (1996).

Table 2.1 Availability of panel data in EU-15

Country	ECHP Data Employed
Germany	1994-1997 inclusive
Denmark	1994-1997 inclusive
Netherlands	1994-1997 inclusive
Belgium	1994-1997 inclusive
Luxembourg	1994-1997 inclusive
France	1994-1997 inclusive
United Kingdom	1994-1997 inclusive
Ireland	1994-1997 inclusive
Italy	1994-1997 inclusive
Greece	1994-1997 inclusive
Spain	1994-1997 inclusive
Portugal	1994-1997 inclusive
Austria	1995-1997 inclusive
Finland	1996-1997 inclusive
Sweden	Some data available for 1997

The guidelines imposed by Eurostat regarding the use of the ECHP users' database are strictly adhered to, so that the anonymity and statistical robustness of the datasets are upheld. Due to the late adoption of the ECHP survey by Austria (1995) and Finland (1996) data for these countries are unavailable prior to their joining the ECHP team. Table 2.1 sets out the availability of datasets for each Member State.[6]

Overview of Housing in Europe

This section provides an overview of selected dwelling details for 14 European countries. Data are presented on a year-by-year and country-by-country basis. This section begins with an overview of the number of dwellings and population size in each country under analysis. An outline of the proportions of households in each country living in various basic house types is also provided in this section, together with an analysis of occupancy levels and housing tenure. Much of these data have not been analysed on a pan-European level hitherto because of the unavailability of comparable data.

Number of Dwellings, Population Size and Mean Occupancy

From Table 2.2 it can be seen that the largest number of dwellings in Europe are found in Germany (36.9 million), a country with also the highest population (79.8 million). Italy has a very large population (56.8 million), but does not have among the highest number of dwellings per country (12.9 million), and is found to have the highest occupancy level by far across this group of European countries, with a mean occupancy of 4.4 persons per dwelling. However, the large populations of France (56.7 million) and the UK (56.5 million) are matched with high numbers of dwellings respectively (27.7 million and 23.8 million). Conversely, Luxembourg has the smallest population in Europe (under 400,000) and the lowest number of dwellings (145,000). This is followed by Ireland, a country marked by a relatively small population (3.6 million) and relatively low numbers of dwellings (just over 1.1 million). However, only Italy has a higher mean occupancy per dwelling; Ireland's rate of 3.3 persons per dwelling is, as such, notable.

Table 2.2 Population levels (millions) (P), number of dwellings (millions) (D) and mean occupancy (persons per household) (O), 1996

	D	DK	NL	B	L	F	UK	IRL	I	EL	E	P	A	FIN	S
P	79.8	5.1	15.0	10.0	0.4	56.7	56.5	3.6	56.8	10.3	38.9	9.9	7.8	5.0	8.6
D	36.9	2.4	6.5	4.1	0.1	27.7	23.8	1.1	12.9	4.0	13.6	3.1	3.6	2.2	4.9
O	2.2	2.1	2.3	2.4	2.7	2.0	2.4	3.3	4.4	2.6	2.9	3.2	2.2	2.3	1.8

Source: Derived from Eurostat (1996).

---

[6] It should be noted that some data from the ECHP were either unavailable or statistically insignificant and have been omitted; such omissions are marked '–'.

Dwelling Type

An overview of the various house types found across Europe is now provided (Table 2.3). These are comparable data which are useful in delineating the growth and development of various housing stocks across Europe.

Table 2.3 Dwelling types in Europe (mean, 1994-97): % and number of households (millions)

	% Detached	No. Detached	% Semi-detach./ Terraced	No. Semi-detached/ Terraced	% Small MFDs	No. Small MFDs	% Large MFDs	No. Large MFDs
D	26.0	9.6	13.4	4.9	34.3	12.7	18.8	6.9
DK	50.2	1.2	12.6	0.3	12.7	0.3	19.2	0.5
NL	13.0	0.8	51.8	3.4	4.4	0.3	18.1	1.2
B	33.9	1.4	42.0	1.7	11.1	0.5	7.7	0.3
L	38.1	0.0	30.8	0.0	19.1	0.0	10.1	0.0
F	38.6	10.7	21.4	5.9	13.3	3.7	24.5	6.8
UK	22.5	5.4	58.8	14.0	11.3	2.7	3.9	0.9
IRL	51.0	0.6	42.9	0.5	2.3	0.0	0.9	0.0
I	20.2	2.6	12.1	1.6	37.3	4.8	23.5	3.0
EL	31.8	1.3	19.0	0.8	29.3	1.2	18.4	0.7
E	17.5	2.4	19.2	2.6	18.8	2.6	43.8	6.0
P	51.1	1.6	26.7	0.8	13.8	0.4	4.9	0.2
A	47.9	1.7	4.5	0.2	11.8	0.4	27.6	1.0
FIN	47.2	1.0	17.6	0.4	2.4	0.1	30.0	0.7
EU14	34.9	2.9	26.6	2.7	15.9	2.1	18.0	2.0
EU10	36.8	3.2	29.6	3.1	12.3	2.1	16.1	1.8

Note: MFD = Multi-family dwellings.

Table 2.3 reports the proportionate breakdown of dwelling type across Europe. The data demonstrate that an interesting, highly varied pattern of house type is found across Europe. The absolute numbers to which these proportions correspond is also provided. The most interesting results relate to the dramatically differing levels of small and large multi-family dwellings across the Member States.

Detached Detached homes are, by far, the most preponderant housing type in Denmark, Ireland and Portugal, where over half of all houses are of this variety. However, in absolute numbers, the highest levels of detached dwellings are found in France (10.7m) and Germany (9.6m). Only 13% of Dutch households are detached and the corresponding proportion is 18% for Spanish housing, though there appears to be an upward trend for this housing type in Spain. Across EU-14, the average proportion of the total housing stock which is detached is calculated to be 36%.

Semi-detached and Terraced Semi-detached and terraced homes are most popular in the UK (59% of the total stock), followed by the Netherlands (52%) and Ireland (43%). The largest number of semi-detached and terraced housing is also found in the UK, with some 14 million such house types. There are virtually no homes of this variety in Austria (4%). Exactly a quarter of all housing in Europe is semi-detached or terraced.

Apartments Small multi-family dwellings (MFDs) – defined as apartments in small complexes (less than 10 units) – are common in Italy (37% of dwelling stock), Germany (34%) and Greece (29%), however there are virtually no such dwellings in Ireland or Finland (2% respectively), and the proportion of housing set out as apartments in the Netherlands is similarly low (4%). In actual numbers, the highest levels of small MFDs are found in Germany (12.7 million) and Italy (4.8 million). Across Europe, it is calculated that 15% of all housing are small multi-family dwellings. Large MFDs – defined as apartments in large complexes (10 or more units) – are a popular form of housing construction in Spain (44%) and Finland (30%), however, once more, accommodation of this type is highly rare in Ireland (1%), the UK (4%) and Portugal (5%). The highest number of large MFDs is found in Germany (6.9 million) and France (6.8 million). Over the period 1994-97, 19% of housing in Europe was of the large multi-family dwelling type.

Housing Tenure

It is useful to identify levels of housing tenures across Europe, as this plays an important role in the interpretation of the results on housing conditions, discussed later. Owner-occupiers can be regarded as fully autonomous, and, as such, may be more likely to invest in remedial work than tenant householders who may feel neither responsible nor authorised to do so. There are three categories of housing tenure: owner-occupiers, rent-paying tenants and rent-free tenants (i.e. those to whom accommodation is provided by the State free of charge). A breakdown of social and private tenants is provided in Chapter 6. As with the previous section, each household type is dealt with sequentially.

Owners Owner-occupier households are generally the most common housing tenure across Europe, as Table 2.3 reports, with 67.1% of households on aggregate in this category. However, the more affluent, industrialised countries tend to have lower levels of owner-occupier households. The highest proportion of such households is found in Ireland (86%), Spain (81%) and Greece (79%). Germany has the lowest level of home-ownership in the group (42%) and is the only country that has more tenant households (54%) than owners, though Germany has among the highest levels of owner-occupiers in absolute numbers (15.6 million). In addition, the Netherlands demonstrates relatively low proportions of owner-occupier households (52%).

Table 2.4 Housing tenure in Europe (mean, 1994-97): % and number of households (millions)

	% Owner-occupier	No. Owner-occupier	% Tenants	No. Tenants	% Rent-free	No. Rent-free
D	42.3	15.6	54.4	20.1	3.2	1.2
DK	62.5	1.5	37.0	0.9	0.5	0.0
NL	52.1	3.4	47.0	3.1	1.0	0.1
B	67.7	2.8	28.9	1.2	3.4	0.1
L	72.2	0.1	24.7	0.0	3.1	0.0
F	56.5	15.7	38.1	10.6	5.4	1.5
UK	68.4	16.3	28.3	6.7	1.6	0.4
IRL	85.8	0.9	13.3	0.1	1.8	0.0
I	74.0	9.5	19.1	2.5	6.9	0.9
EL	78.7	3.1	17.9	0.7	3.4	0.1
E	80.7	11.0	12.9	1.8	6.4	0.9
P	68.5	2.1	20.0	0.6	11.5	0.4
A	58.7	2.1	34.6	1.2	6.6	0.2
FIN	73.8	1.6	16.9	0.4	9.2	0.2
EU14	67.3	6.1	28.1	3.6	4.6	0.4
EU10	64.0	6.0	32.3	4.4	3.6	0.4

Tenants Rent-paying tenants are also most commonly found in Germany (54%), the Netherlands (47%) and France (38%), but not in Spain or Ireland (13% respectively) or Finland (17%). Germany (20.1 million) and France (10.6 million) have the highest numbers of such housing tenure. Across this set of European countries, 28% of households fall into this category of housing tenure (Table 2.4).

Rent-free This category of housing encompasses local-authority tenants who pay no rent for their dwelling, i.e. the State subvents their accommodation costs in full. From Table 2.4 it can be seen that housing provided rent-free is the least common form of housing tenure across Europe, with less than 5% of all households in this tenure category. Such accommodation is hardly evident at all in Denmark or the Netherlands (1% respectively) or the UK and Ireland (2% respectively). While such accommodation is not very frequently found in European countries, Portugal and Finland have moderate levels of such household tenure (12% and 9% respectively). In absolute terms, the highest levels of dwellings provided rent-free by the State include 1.5 million in France and 1.2 million such dwellings in Germany.

Household Size (Occupancy)

Table 2.5 Occupancy levels (mean persons per household, 1994-97): % and number of households (millions)

	D	DK	NL	B	L	F	UK	IRL	I	EL	E	P	A	FIN	S	EU-15
%1	23	32	26	25	22	25	26	14	15	17	14	15	22	21	33	22
#1	9	1	2	1	0	7	6	0	2	1	2	1	1	1	2	2
%2	36	33	34	30	28	32	34	23	23	27	25	27	30	32	34	30
#2	13	1	2	1	0	9	8	0	3	1	3	1	1	1	2	3
%3	18	15	14	18	18	17	16	16	24	20	21	23	16	18	13	18
#3	7	0	1	1	0	5	4	0	3	1	3	1	1	0	1	2
%4	16	14	18	18	19	16	16	19	25	24	24	20	18	18	14	19
#4	3	0	1	0	0	2	2	0	2	1	2	0	0	0	0	1
%5	5	4	7	7	10	7	5	15	9	8	11	8	8	8	5	8
#5	2	0	1	0	0	2	1	0	1	0	2	0	0	0	0	1
%6	2	2	2	3	3	3	2	13	3	4	6	6	6	3	2	4
#6	1	0	0	0	0	1	1	0	0	0	1	0	0	0	0	0

Household size across Europe is now presented, using 'persons per household' as a measurement. The averages of the four years of data are compared across all household occupancy levels (one- to six-person households). Table 2.5 presents a useful overview of housing composition across EU-15. In doing this, it becomes clear that the two-person household is the most dominant household size in each of the 14 EU countries under analysis except Italy (30% of all households are of this composition). The one-person household is also dominant (22% across EU-14). On a country-by-country examination of the data, it becomes apparent that Sweden and Denmark have the highest incidences of small (one-person) occupied households, with 33% and 32% respectively, although in actual numbers Germany and France have the highest levels of one-person households, with some 8.5 million and 6.9 million respectively. Conversely, Ireland has the largest proportion of large households across EU-14, with 13% of all households containing six or more persons, while France and Spain have the highest absolute numbers of such households (with 800,000 households in each country with six or more persons).

Housing Conditions in Europe

A number of social indicators of housing conditions pertaining to thermal efficiency are reported across EU-14 using the longitudinal ECHP datasets. It should be remembered that these data are self-reporting and are, thus, open to a margin of error associated with such data. However, much of the data are objective (as opposed to subjective) indicators of housing conditions relating to the presence or absence of certain household characteristics; as such, this reduces the amount of error associated with self-reporting data. Nonetheless, readers are cautioned with regard to findings based on subjective self-reported data, as cultural differences can result in substantial variations across multi-country results and differing interpretations of the same data.

Leaking Roofs

A leaking roof has a number of adverse impacts on households' well-being. Besides the obvious implications for energy efficiency (and the associated excess fuel bills), a leaking roof may cause damp and mould spores to develop in the dwellings' walls. Such spores are pernicious to human health, especially for the very young and the elderly, and can lead to serious respiratory conditions, such as bronchitis and asthma (March *et al.*, 1999). Therefore, it is disturbing to find in this research that 16% of households in Greece and 19% of Portuguese households are suffering from leaking roofs. The situation is also unsatisfactory in Spain, with 11% (or 1.6 million households) affected. Germany demonstrates the joint-highest numbers of households affected by this condition, with 1.6 million dwellings affected by leaking roofs.

Table 2.6 Households with leaking roofs (1994-97): % and number of households affected ('000)

	% 1994	% 1995	% 1996	% 1997	% Mean	No. Mean
D	5.5	4.2	3.1	-	4.3	1,590
DK	4.1	4.1	4.0	3.5	3.9	90
NL	5.6	5.3	3.6	3.2	4.4	290
B	6.3	6.8	4.7	4.8	5.7	230
L	5.1	6.0	3.9	-	5.0	10
F	6.1	5.8	4.9	4.5	5.3	1,470
UK	4.4	4.2	3.3	3.4	3.8	900
IRL	4.7	4.2	3.8	3.8	4.1	50
I	8.1	6.2	5.4	5.3	6.3	810
EL	18.2	16.3	16.0	15.1	16.4	660
E	12.7	9.9	11.2	11.8	11.4	1,550
P	20.0	18.0	17.8	18.9	18.7	580
A	-	4.4	3.3	3.0	3.6	130
FIN	-	-	2.6	3.1	2.9	60
EU14	8.4	7.3	6.3	6.7	6.8	601.4
EU10	5.2	5.0	3.7	3.7	4.3	482.0

Furthermore, France displays relatively high numbers of households with this condition, with 1.5 million households suffering. Leaking roofs do not appear to be common throughout the rest of Europe and an EU-14 average of less than 7% of all households is found. Longitudinally, the data are interesting as they show that leaking roofs are becoming somewhat less of a problem for almost all EU countries in this analysis (Table 2.6). This is likely to be related to improving living standards over time.

Hot Running Water

The provision of a hot-water system would be considered by most as an essential household attribute, yet (for the survey period) 70% of Greek households did not have this basic household amenity; this represents 2.8 million households in Greece. Portugal also demonstrates a significant proportion of households lacking hot-water facilities (23%), though this amounts to just 700,000 households.

The vast majority of all other European households do not lack hot running water. Because of the large numbers of dwellings in Germany, small proportions of households can amount to significant numbers in absolute terms, and the relatively small proportion of households lacking hot running water (4.6%) amounts to some 1.7 million households. With the exception of Italy and Greece, all other EU-14 countries included in this component of the ECHP are demonstrating significantly reduced problems in this domain over the period 1994-97 (Table 2.7). Again, increased economic prosperity is likely to play a substantial part in this improvement over time.

Table 2.7 Households lacking hot running water (1994-97): % and number of households affected ('000)

	% 1994	% 1995	% 1996	% 1997	% Mean	No. Mean
D	6.7	5.4	3.5	2.9	4.6	1,700
DK	1.0	0.8	0.7	0.5	0.8	20
NL	0.7	0.5	0.5	0.3	0.5	30
B	4.4	3.7	3.5	2.9	3.6	150
L	4.0	2.8	2.4	-	3.1	-
F	2.4	2.0	1.9	1.8	2.0	550
UK	0.3	0.2	0.1	-	0.2	50
IRL	4.3	4.3	4.2	3.3	4.0	40
I	2.6	2.5	3.1	2.9	2.8	360
EL	-	71.1	68.7	70.7	70.2	2,800
E	4.5	3.6	3.0	2.8	3.5	480
P	25.3	24.6	21.5	19.1	22.6	700
A	-	2.7	2.5	1.9	2.4	90
FIN	-	-	1.9	1.7	1.8	40
EU14	5.1	9.6	8.4	9.2	8.7	539.2
EU10	3.0	2.5	2.1	1.9	2.3	296.7

Objective Overcrowding

Household overcrowding is considered undesirable for human health, as it has been shown to be linked with increased rates of various viral and bacterial infections, especially in the respiratory tract (Marsh *et al.*, 1999). Thus, the levels of overcrowded households may be considered to be important indicators of housing conditions and, more generally, quality of life across EU-14. Objective overcrowding is calculated by dividing the total number of persons in the household by the number of inhabitable rooms in the dwelling (excluding kitchens, bathrooms, halls, garages, etc.). A household with more than one person per room is considered 'objectively overcrowded' – a standard definition of overcrowding (Eurostat, 1996).

The worst levels of overcrowding are found in southern Europe, with Italy (24%), Portugal (20%) and Spain (16%) demonstrating the most objectively overcrowded households. In northern Europe, Ireland (15%) is demonstrating high levels of objective overcrowding, while Finnish households also fare badly (13%) using the Eurostat calculation. With the exceptions of Germany, Denmark and Spain, all other countries in this pool are showing decreased levels of objective overcrowding over time (Table 2.8). Approximately 3.5 million German households are calculated as being objectively overcrowded, followed by 3.1 million households in Italy and 2.3 million households in France.

Table 2.8 Households objectively overcrowded (1994-97): % and number of households affected ('000)

	% 1994	% 1995	% 1996	% 1997	% Mean	No. Mean
D	9.4	8.5	7.4	12.4	9.4	3,470
DK	4.1	4.2	7.2	4.4	5.0	120
NL	-	-	-	-	-	-
B	5.0	5.3	5.1	4.6	5.0	210
L	8.7	8.8	7.8	-	8.4	10
F	8.6	8.3	8.2	7.6	8.2	2,270
UK	5.6	5.4	5.1	4.8	5.2	1,240
IRL	16.4	15.7	14.1	12.8	14.8	160
I	25.2	24.7	24.0	23.3	24.3	3,130
EL	-	-	-	-	-	-
E	14.9	16.0	16.7	15.4	15.8	2,150
P	20.9	19.9	19.6	19.2	19.9	620
A	-	-	-	-	-	-
FIN	-	-	14.2	12.3	13.3	290
EU14	11.9	11.7	11.8	11.7	11.8	1242.7
EU10	8.3	8.0	8.6	8.4	8.7	971.3

Subjective Overcrowding (Self-Reporting)

Table 2.9 Households subjectively overcrowded (1994-97): % and number of households affected ('000)

	% 1994	% 1995	% 1996	% 1997	% Mean	No. Mean
D	16.2	12.7	11.6	19.5	15.0	5,535
DK	16.1	16.2	15.6	15.9	16.0	384
NL	9.9	9.4	10.3	9.6	9.8	637
B	15.2	14.5	14.7	13.8	14.6	599
L	13.0	11.6	9.8	-	11.5	12
F	15.8	13.5	12.1	13.0	13.6	3,767
UK	20.9	20.6	19.4	11.2	18.0	4,284
IRL	15.3	12.4	11.8	10.5	12.5	138
I	22.2	19.5	18.5	19.4	19.9	2,567
EL	31.7	27.2	24.7	23.9	26.9	1,076
E	23.0	21.8	21.6	20.1	21.6	2,938
P	31.8	28.2	26.1	25.6	27.9	865
A	-	15.7	14.4	12.2	14.1	508
FIN	-	-	15.2	15.4	15.3	337
EU14	19.3	17.2	16.1	16.2	16.9	1689.1
EU10	15.3	14.1	13.5	13.5	14.0	1620.1

In this section of the survey, households are asked to declare whether or not they have enough space to meet their needs. It is, thus, a subjective indicator of fuel poverty and allows for cross-checking of results with the objective (calculated) measure of overcrowding, reported in a previous section. An EU-14 average level of 16% is found, and southern countries like Italy (20%), Spain (22%), Greece (27%) and Portugal (28%) all exhibit very high levels of overcrowding. Turning the analysis solely to northern Europe, an average rate of overcrowding of 14% is reported, with high levels found in Denmark (16%) and the UK (18%). The UK figure for self-reported overcrowding is far higher than that found for objective overcrowding, so a large mismatch is reported here. Germany appears to have the highest absolute number of self-reported overcrowded houses (with 5.5 million affected). These figures cross-check well with the results for objective overcrowding. Generally, subjective indicators of overcrowding provide more generous, upper bound results, i.e. households tend to declare a shortage of space even when they are not 'overcrowded' using a quantitative, technical measurement.[7]

There does not appear to be a distinct trend developing across EU-14 with regard to the levels of subjective overcrowding. While slight reductions are reported in Denmark and Austria, and substantial reductions are reported in Ireland

---

[7] Results for 'objective' overcrowding demonstrated that only Ireland and Italy break this pattern, i.e. Irish and Italian householders under-declare their objective level of overcrowding.

and the UK (decreases of 46% in the UK and 31% in Ireland), other countries are providing mixed self-reported data over time (Table 2.9).

Presence of Damp Walls and/or Floors

The presence of damp indicates that the dwelling is not energy efficient. It may also be a manifestation of a continuously unheated or ineffectively heated home. In both cases, it acts as a good objective indicator of fuel poverty. Households are assessed to identify if there are any patches of damp on either the walls, floors or foundations in their home. Some 12.7% of all European households contain damp patches. Again, Greece (18.8%), Spain (21.5%) and Portugal (33.4%) are suffering worst in this regard (Table 2.10).

In northern Europe, an average incidence of 9.8% is found, and the UK (13%), Belgium (14.4%) and France (16.3%) all perform poorly with regard to the presence of household damp. These results are particularly important from a public-health perspective, as chronic exposure to damp is strongly associated with increased rates of respiratory disease in humans (Dales *et al.*, 1991). In physical numbers, the largest levels of damp are found in France (4.5 million) and the UK (3.1 million). There have been some quite dramatic reductions in the presence of damp throughout the mid-1990s. The UK has experienced a 53% fall in damp spores, while the level of household damp has fallen by 44% in Italy and by 28% in France.

Table 2.10 Damp households (1994-97): % and number of households affected ('000)

	% 1994	% 1995	% 1996	% 1997	% Mean	No. Mean
D	9.5	7.8	6.4	-	7.9	2,915
DK	6.6	6.2	6.5	5.6	6.2	149
NL	12.0	12.0	9.8	9.5	10.8	702
B	15.8	16.7	12.3	12.8	14.4	590
L	7.2	8.2	7.2	-	7.5	8
F	19.5	17.1	14.6	14.0	16.3	4,515
UK	17.2	14.3	12.2	8.1	13.0	3,094
IRL	10.5	9.4	8.9	9.4	9.6	106
I	7.2	5.4	4.8	4.1	5.4	697
EL	20.8	17.7	18.5	18.2	18.8	752
E	25.6	19.2	20.4	20.8	21.5	2,924
P	32.7	32.3	33.5	35.2	33.4	1,035
A	-	10.1	8.3	8.1	8.8	317
FIN	-	-	3.9	3.7	3.8	84
EU14	15.4	13.6	12.0	12.5	12.7	1277.7
EU10	12.3	11.3	9.0	8.9	9.8	1248.0

Rotten Window Frames

In this part of the ECHP, respondent households are asked to check their windows for condition. Window frames which have become rotten are not energy efficient and, as such, can be considered a good (objective) indicator of fuel poverty. Rot is most commonly found in Portugal where a quarter (25.2%) of all households have rotten windows, compared with an EU-14 (all countries) mean of 8.6%. In northern Europe 12.7% of British households are suffering from this adverse housing condition, and in France the corresponding percentage is 10.4% (Table 2.11). This compares with an EU-10 (northern European) mean of 7.2%. Rot is most commonly found, in absolute terms, in the UK (some 3 million households), followed by France (2.9 million). There have been very substantial decreases in the proportion of households with rotten windows over the time period under analysis. The reduction in the UK represents a 35% fall on 1994 levels, while in Ireland the corresponding fall is 25%. In southern Europe, Greece has reduced its level of rotten windows by 37%.

Table 2.11 Households with rotten window frames (1994-97): % and number of households affected ('000)

	% 1994	% 1995	% 1996	% 1997	% Mean	No. Mean
D	6.4	5.4	4.2	-	5.3	1,956
DK	6.1	5.2	5.8	5.2	5.6	134
NL	9.8	10.2	9.8	8.8	9.7	631
B	9.2	10.3	8.7	8.3	9.1	373
L	4.8	5.2	4.4	-	4.8	5
F	11.2	10.7	9.7	10.0	10.4	2,881
UK	15.3	14.1	11.6	9.9	12.7	3,023
IRL	8.9	6.4	7.0	6.7	7.3	80
I	8.0	6.2	5.2	5.3	6.2	800
EL	11.8	9.6	8.5	7.4	9.3	372
E	9.7	7.7	6.4	6.6	7.6	1,034
P	24.2	26.1	25.0	25.3	25.2	781
A	-	5.5	4.4	3.8	4.6	166
FIN	-	-	2.5	2.6	2.6	57
EU14	10.5	9.4	8.1	8.3	8.6	878.1
EU10	9.0	8.1	6.8	6.9	7.2	930.6

Domestic Energy-Efficiency Levels[8]

Energy-efficiency standards in housing vary dramatically across Europe. As was discussed in the introduction, certain colder climates need better protection from the elements, however all parts of Europe experience cold winter spells; even the

---

[8] The results in this and the next section come from Eurostat (1999) and are not calculated using the ECHP datasets, as the ECHP does not contain such variables.

relatively warm climates of Spain and Italy endure average January temperatures of about 6°C. From the data analysed in this chapter, countries in southern Europe appear to be the most energy inefficient when examined using multiple criteria. However, data on energy-efficiency standards in southern Europe are notoriously difficult to obtain, and their reliability is often suspect. Eurostat (1999) have provided data for Greece but only scant data for Portugal and no data whatsoever for Italy or Spain. However, the data that is available, together with the analysis of other housing indicators in this chapter, indicate that energy efficiency in southern-European housing is not a priority in their building regulations.

From Table 2.12 it can be seen that, with the exception of roof insulation, Ireland performs considerably poorer than its northern-European counterparts, on average, with regard to the penetration of various domestic energy-efficiency measures. Ireland has the lowest proportion of double-glazed dwellings in northern Europe – only one-in-three Irish households are double-glazed compared with an average of over three-in-five in northern Europe. The UK fares poorly with regard to its levels of wall insulation; only a quarter of all houses in 1996 are equipped with this measure. The level of floor insulation in the UK, at 4%, is the lowest in this group of 13 European countries.

Table 2.12 Percent of households with various energy-efficiency measures (1996)

	Cavity-wall insulation (%)	Double-glazing (%)	Floor insulation (%)	Roof insulation (%)
D	24	88	15	42
DK	65	91	63	76
NL	47	78	27	53
B	42	62	12	43
F	68	52	24	71
UK	25	61	4	90
IRL	42	33	22	72
EL	12	8	6	16
P	6	3	2	6
A	26	53	11	37
FIN	100	100	100	100
S	100	100	100	100
NOR	85	98	88	77
Mean	49	61	37	63

Elsewhere in northern Europe, Austria performs poorly with regard to its domestic energy-efficiency standards. Only one-in-nine Austrian households (11%) have floor insulation and only a quarter (26%) have cavity-wall insulation. Belgium also has a relatively energy inefficient dwelling stock, with just 12% of dwellings equipped with floor insulation and a third fitted with roof insulation. On the other hand, countries like Finland, Norway and Sweden have exemplary levels of residential energy efficiency, with almost all homes fitted with all four key energy-saving measures. Denmark also performs well with regard to the measures examined in this analysis. It should be noted that, because of variations in housing types across Europe, certain housing styles are not amenable to certain forms of retrofitting. For instance, it is far more practicable to retrofit wall insulation into a 1960s cavity-wall dwelling than a 19th century Victorian dwelling.

These results have strong implications for policy-makers. Energy inefficiency in the home can make home-heating such a burden to low-income householders that they cannot afford to heat their home adequately and, as such, live in fuel poverty (Boardman, 1991). The implications on human health of living in fuel-poor conditions are particularly pernicious and have been shown to be associated with higher incidences of various cardiovascular and respiratory disorders (Rudge and Nicol, 2000). In addition, chronic, long-term exposure to cold and damp in the home has been associated with premature mortality (Gemmell *et al.*, 2000).

Domestic Energy Consumption

Countries demonstrating poor levels of domestic energy efficiency are consuming greater amounts of energy than necessary, as those inhabiting inefficient dwellings must consume more fuel to heat their homes adequately. As can be seen from Figure 2.1, Irish energy consumption per household is the highest in Europe, at 102,000 Megajoules per annum, compared to an EU-14 average of 74,000 MJ.

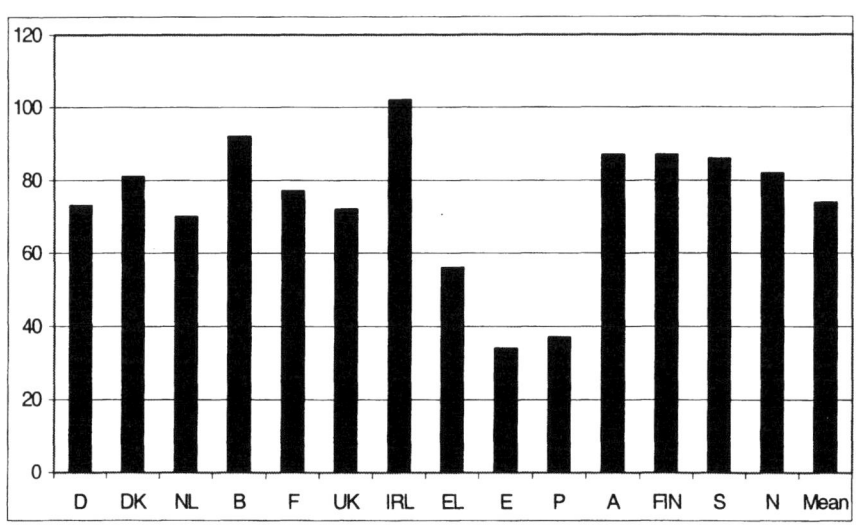

Figure 2.1 Energy consumption per household ('000 MJ, 1996)

It is important to state that factors such as aggregate household composition and occupancy levels play a part in domestic energy use. If the above data were adjusted to a per-capita basis, the relative ranking would change (see Table 2.4). However, it is likely to be of little coincidence that the Irish housing stock is among the least energy efficient in northern Europe. Consequently, environmental emissions, such as CO_2, SO_2, NO_x and particulate matter are also greater. This is of considerable importance given that some European countries – including Ireland – are having extreme difficulty in meeting their agreed targets for stabilisation of greenhouse-gas emissions under the Kyoto Protocol and acidification precursors under the Gothenburg Agreement. The relatively low levels of household energy consumption in Greece, Spain and Portugal are due, to a large extent, to the poorer living standards generally found in these nations.

Housing Finances in Europe

The next set of indicators pertain to affordability of housing and levels of financial hardship in meeting housing costs across EU-14. Affordability is generally considered fundamental to housing satisfaction, and there is a considerable literature demonstrating the relationship between housing finances and levels of satisfaction using mainly self-reported survey data. This research analyses three questions contained in the ECHP survey. The first asks householders to consider their ability to meet housing costs; the second asks households whether they can afford adequate home heating, or are otherwise fuel-poor; the third asks households about their ability to meet scheduled utility bills on time. Furthermore, an additional question is asked to tenant householders only, which assesses levels of State fuel subsidisation across Europe.

Housing Costs Are a Financial Burden

Households were asked to consider their income and their expenditure on housing, i.e. their mortgage or rent payments, and declare whether or not they found these costs to be financially burdensome. Results show that, while the less-affluent southern European countries like Spain, Greece and Portugal are all demonstrating relatively high levels of financial hardship in this regard, richer (mainly northern) European countries with relatively high standards of living, such as Italy, Belgium, Ireland and the UK, are also exhibiting alarming difficulties in meeting their mortgage and rent payments.

A quarter of households in Belgium and over a third of Italian households are declaring housing costs as burdensome. In Ireland, one-in-five households are finding it difficult to meet housing costs, while the respective proportion in both the UK and France is 17%. France, Germany and Italy demonstrate the highest absolute levels of financial burden, with over four-and-a-half million households affected in each country. Conversely, such costs are not problematic in the Netherlands and Luxembourg. Upon examining the time-series data, it becomes apparent that some countries (Denmark, Ireland, the UK and Spain) are realising successive reductions in their respective levels of financial burden over the period 1994-97. However, the results also show that the converse is true for Belgium and Portugal (Table 2.13). It is important to note that, for many indicators, the mean figures often give an inaccurate picture of the current situation. This is especially true for countries like the UK which have experienced very significant reductions in housing deprivation indicators over the time series analysed.

Table 2.13 Households whose housing costs are a "heavy" financial burden (1994-97): % and number of households affected ('000)

	% 1994	% 1995	% 1996	% 1997	% Mean	No. Mean
D	12.4	10.5	10.3	16.8	12.5	4,613
DK	7.6	6.4	7.4	5.8	6.8	163
NL	4.6	4.9	5.1	4.6	4.8	312
B	22.6	24.4	25.8	26.6	24.9	1,021
L	5.6	5.0	5.1	-	5.2	5
F	17.7	17.2	17.8	16.6	17.3	4,792
UK	22.2	21.3	18.3	6.9	17.2	4,094
IRL	22.1	19.9	19.5	16.4	19.5	215
I	33.6	34.0	35.7	37.0	35.1	4,528
EL	31.0	24.8	23.6	21.1	25.1	1,004
E	35.5	33.2	29.6	29.2	31.9	4,338
P	19.4	18.6	18.8	21.1	19.5	605
A	-	11.5	10.2	10.5	10.7	385
FIN	-	-	13.7	15.3	14.5	319
EU14	19.5	17.8	17.2	17.5	17.5	1885.3
EU10	14.4	13.5	13.3	13.3	13.3	1591.9

Unable to Afford Adequate Home Heat

The ability to heat the home adequately is a fundamental aspiration. Households which declare an inability to do this may be considered fuel-poor, for this is a classic qualitative definition of fuel poverty (Lewis, 1982). Results show that an alarming 45.5% of households in Greece, 54.9% in Spain and 74.4% in Portugal declare this inability (Table 2.14). The variation of results across countries is dramatic: households in Germany, the Netherlands, Austria, Luxembourg and Denmark appear to have no significant problems heating their homes (all report incidences of 3% or less respectively), while some relatively rich nations, like Italy, report very substantial levels of fuel poverty (21.5%). The average rate throughout all 14 EU countries analysed is 16.9%, but it is useful to analyse a sub-section of these countries separately.

If the analysis is narrowed to the 10 northern or non-Mediterranean countries, a substantially different mean rate of fuel poverty is found using this indicator: a much lower rate of just 4.0%. In this instance, countries such as the UK (5.8%) and Ireland (6.4%) are identifiable as relatively high sufferers of this indicator of fuel poverty. The highest levels of fuel-poor households in absolute numbers are found in Spain, (7.4 million), Italy (2.8 million) and Portugal (2.3 million households). Looking at the longitudinal results, it is worthwhile to note that, while many countries are realising significant reductions in fuel poverty across the time series, the most notable aggregate decreases in the incidence of fuel poverty are by the UK (a reduction of 70%), Denmark (38%) and Ireland (36%). It is unlikely to be a coincidence that all three countries have relatively aggressive State-funded fuel-poverty strategies in place.

Table 2.14 Households unable to heat home adequately (1994-97): % and number of households affected ('000)

	% 1994	% 1995	% 1996	% 1997	% Mean	No. Mean
D	2.0	1.5	1.4	-	1.6	590
DK	4.2	2.9	2.8	2.6	3.1	74
NL	2.0	1.8	2.0	2.2	2.0	130
B	4.6	4.1	2.8	3.0	3.6	148
L	2.6	3.1	3.5	-	3.1	3
F	8.5	7.3	7.0	5.8	7.2	1,994
UK	8.9	6.2	5.3	2.7	5.8	1,380
IRL	8.0	5.9	6.5	5.1	6.4	70
I	22.4	22.7	20.6	20.3	21.5	2,774
EL	46.8	45.5	46.8	42.9	45.5	1,820
E	58.7	57.7	53.3	49.7	54.9	7,466
P	75.8	74.9	73.8	72.9	74.4	2,306
A	-	2.5	1.9	1.8	2.1	76
FIN	-	-	4.7	4.7	4.7	103
EU14	20.4	18.2	16.6	17.8	16.9	1352.4
EU10	5.1	3.9	3.8	3.5	4.0	456.8

Households Declaring an Inability to Pay Utility Bills

A household which has been unable to pay on time a scheduled utility (gas or electric) bill over the previous 12 months is most likely finding it difficult to keep the home adequately heated and, as such, this affordability indicator also assists in identifying fuel poverty. Those who were unable to keep up to date with their utility bills may also have suffered disconnection from the supplier, compounding the experience of fuel poverty. The results of this section demonstrate that Mediterranean nations do not appear to suffer higher than average from an inability to pay utility bills on time. In fact, Italy, Spain and Portugal rank among the best countries in this exercise, hence this result is something of an anomaly. While the highest level of late-payment of utility bills is found in Greece, where a remarkable 32% are affected (Table 2.15), the UK (8.1%) and France (7.6%) also perform poorly with this indicator. The mean (EU-14) proportion of households unable to pay their utility bills on time is 6.2%, while the corresponding statistic for the EU-

10 group of northern-European countries is 4.4%. Some 2.1 million French households, over 1.9 million British households and 1.3 million Greek households are reporting this social indicator. The time-series shows that the UK has reduced the proportion of households unable to pay utility bills on time by a quarter, while Ireland has achieved an even larger reduction of 42%. Spain also reports positive longitudinal results with a reduction of 21% over four years. Other countries report less clear results, while some are showing increased difficulties in meeting the costs of utility bills over time.

Table 2.15 Households unable to pay scheduled utility bills (1994-97): % and number of households affected ('000)

	% 1994	% 1995	% 1996	% 1997	% Mean	No. Mean
D	2.2	1.6	1.5	-	1.8	664
DK	3.2	2.6	2.4	2.6	2.7	65
NL	1.4	1.1	1.2	1.4	1.3	85
B	7.7	6.1	6.9	6.7	6.9	283
L	3.1	2.2	2.8		2.7	3
F	8.8	7.2	7.3	7.0	7.6	2,105
UK	9.4	7.8	7.0	-	8.1	1,928
IRL	8.4	6.3	6.1	4.9	6.4	70
I	3.8	3.8	4.5	4.1	4.1	529
EL	36.5	30.1	-	29.3	32.0	1,280
E	5.2	4.3	3.7	4.1	4.3	585
P	3.1	1.7	1.7	1.8	2.1	65
A	-	1.1	1.1	1.2	1.1	40
FIN	-	-	11.4	0	5.7	125
EU14	7.7	5.8	4.4	5.7	6.2	559.1
EU10	5.5	4.0	4.8	3.4	4.4	536.8

Tenant Households Paying for Home-Heating in Addition to Rent

This section identifies the proportions of tenant households per country that pay for their home-heating requirements in addition to their rent. This is useful, as it identifies broadly the levels of fuel subsidisation across Europe, and assists in identifying the levels of hardship facing those households suffering from fuel poverty, and whether these levels of subsidisation are influencing positively the levels of fuel poverty. Table 2.16 illustrates the results of this exercise.

There is a highly varied level of fuel subsidisation across EU-14. Finland has the smallest proportion of tenant households paying for their home-heating, with just 7% having to pay in addition to their rent; conversely, 97% of Greek rent-paying tenants and 91% of British tenant households pay for their home heating, while similarly high levels are found in Ireland (88%). In actual numbers, Germany has the highest level of tenant households paying for their heat in addition to their rent, with 9 million households paying for home heating. Similarly high levels are found in France (6.4 million) and the UK (6.1 million). With the notable exception of Portugal (where there has been a 150% increase), most countries are not displaying either a significant upward or downward trend in the level of tenant households paying for their home-heating requirements in addition to their rent. While these results do not show a perfect relationship between the level of fuel poverty and state subsidisation of fuel, it is clear that low levels of aggregate fuel subsidies tend to correlate generally with higher aggregate rates of fuel poverty, as would be predicted.

Table 2.16 Tenant households that pay for home heating in addition to rent (1994-97): % and number of households affected ('000)

	% 1994	% 1995	% 1996	% 1997	% Mean	No. Mean
D	45.8	45.1	44.0	-	45.0	9,033
DK	20.6	20.6	20.9	21.5	20.9	186
NL	81.4	82.1	81.4	82.0	81.7	2,496
B	80.3	79.2	82.2	82.9	81.2	962
L	66.2	66.5	64.9	-	65.9	16
F	58.2	62.1	61.3	62.5	61.0	6,438
UK	88.0	91.5	91.2	91.1	90.5	6,096
IRL	89.8	85.8	88.5	88.9	88.3	129
I	78.9	76.3	76.7	75.8	76.9	1,895
EL	91.4	97.9	98.8	98.7	96.7	692
E	15.6	12.0	16.1	15.5	14.8	260
P	23.8	43.1	51.5	58.2	44.2	274
A	-	76.7	79.7	83.8	80.1	998
FIN	-	-	6.9	7.6	7.3	27
EU14	61.7	64.5	61.7	64.0	61.0	2107.3
EU10	66.3	67.7	62.1	65.0	62.2	2638.1

Housing Satisfaction in Europe

This final section examines self-reported satisfaction with housing. In this section households are asked to rate their satisfaction with their housing on a scale of 1 ("totally satisfied") to 6 ("totally dissatisfied"). The proportion of households who are either "totally dissatisfied" or "very dissatisfied" (5 or 6 ratings) are combined as a measure of housing dissatisfaction, and the results are shown in Table 2.17. These results are considered important, as housing is a fundamental quality-of-life indicator. The link between housing satisfaction and happiness is well-researched and highly correlated; indeed, it has been shown by Barresi et al. (1984), amongst others, to be the key factor influencing happiness, especially in older populations. As with most of the results in this chapter, the analysis is based on self-reporting data, and it is important to stress, once more, the limitations of using such data in cross-country analyses. This is because cultural factors can affect householders' responses, especially in relation to a subjective indicator of happiness like satisfaction with housing. Nonetheless, the results appear to fit very well with the other housing indicators in this analysis, and a clear pattern of housing dissatisfaction is found which corroborates the more objective data on housing conditions presented in previous sections.

Table 2.17 Households "totally" or "very dissatisfied" with housing (1994-97): % and number of households affected ('000)

	% 1994	% 1995	% 1996	% 1997	% Mean	No. Mean
D	6.6	6.3	6.0	-	6.3	2,325
DK	3.8	4.0	3.4	3.9	3.8	91
NL	2.7	2.8	2.6	2.3	2.6	169
B	4.9	5.4	5.2	4.4	5.0	205
L	4.4	3.7	2.9	-	3.7	4
F	4.7	3.8	4.1	3.7	4.1	1,136
UK	8.2	6.8	5.5	-	6.8	1,618
IRL	7.8	5.7	5.7	4.4	5.9	65
I	14.3	11.2	10.0	9.6	11.3	1,458
EL	16.5	12.3	11.5	11.3	12.9	516
E	10.7	8.6	8.1	8.2	8.9	1,210
P	13.2	11.1	9.8	10.9	11.3	350
A	-	-	-	3.7	3.7	133
FIN	-	-	3.5	3.5	3.5	77
EU14	8.2	6.8	6.0	6.0	6.4	668.4
EU10	5.4	4.8	4.3	3.7	4.5	582.3

The findings are again negative for southern Europe, with Greece demonstrating the highest proportion of dissatisfied households, with a level of 13%. Italy and Portugal are also countries which are exhibiting relatively high rates of housing dissatisfaction, with one-in-nine respective households either "totally" or "very" dissatisfied. Spain is also affected (9%) by housing dissatisfaction, with about one-in-eleven households expressing serious dissatisfaction with their housing conditions. Again, Germany shows the highest absolute numbers affected, with 2.3 million households dissatisfied, followed by 1.6 million in the UK and 1.5 million in Italy. Elsewhere, households appear to be relatively content with their housing conditions. As with other self-reported indicators utilised in this chapter, longitudinal trends are often mixed or unclear and certainly difficult to pinpoint with any degree of certainty. However, a few observations can me made. Housing dissatisfaction is falling substantially over the period 1994-97 in Luxembourg (34%), the UK (33%), Ireland (44%), Italy (33%) and Greece (32%). However, other countries are showing either static or somewhat mixed rates of housing dissatisfaction over the years in question (Table 2.17).

These data appear to corroborate the findings regarding the objective housing conditions of the previous sections in this chapter which showed serious problems in southern Europe. It is, however, interesting to note that, despite demonstrating relatively inefficient housing conditions (particularly in relation to energy efficiency) and high levels of overcrowding, households in the UK and Ireland appear relatively content with their housing.

Summary and Conclusions

This chapter has presented an analysis of housing conditions in Europe using the first standardised, longitudinal, pan-European dataset on social indicators. This quantitative analysis of housing conditions (particularly those concerned with energy efficiency), financial situations of householders and levels of satisfaction with housing has shown that there are a number of serious causes for concern. Figure 2.2 illustrates the cumulative results of the 10 key housing indicators presented in this analysis for all 14 countries using the mean unweighted rates produced for each indicator from 1994-97. The figure, although somewhat crude, illustrates clearly the first major cause for concern, namely the poor housing conditions reported in southern Europe. Specifically, there are significant proportions of the Greek, Spanish and Portuguese housing stocks reporting problems with leaking roofs, while large numbers of households in Portugal and Greece do not possess hot-water facilities. All four southern countries demonstrate the highest levels of overcrowding (using both objective and self-reported measurements). Furthermore, damp is found to be a serious problem in Portugal, Spain and Greece. These findings have profound public-health repercussions, as damp, overcrowding and other inadequate housing conditions have been shown to be associated with human health by many epidemiological studies.[9]

To compound these results further, available data for Mediterranean households regarding energy-efficiency levels indicate that these countries suffer from the least energy efficient housing in Europe and the highest levels of fuel poverty. Moreover, the causal link between fuel poverty and excess winter mortality is becoming stronger,[10] and these results seem to corroborate the strong association generally reported; seasonal variations in mortality in southern Europe are several times those found in much of northern Europe (Clinch and Healy, 1999a). It is, perhaps, unsurprising that the highest levels of housing dissatisfaction are reported in these four southern-European nations.

Northern-European countries suffer less from poor housing conditions, however 15% of Irish households are statistically overcrowded and 18% of British households declare a shortage of space. Belgium, France, the UK and Ireland demonstrate the highest levels of poor housing cumulatively using the ten indicators (Figure 2.2). The most notable deficiency regarding housing conditions in northern Europe is specific to the energy efficiency of the dwelling. Both Ireland

---

[9] An enormous, robust epidemiological literature exists in this field. Marsh *et al.* (1999) and Rudge and Nicol (2000) contain some very notable recent findings in this regard.

[10] See, for example, Clinch and Healy (2000c), Eng and Mercer (1998) and McKee *et al.* (1998).

and Britain are found to have thermally inefficient dwelling stocks, the former demonstrating among the lowest levels of cavity-wall insulation, floor insulation and the lowest level of double-glazing in northern Europe. British households are similarly under-insulated, with paltry floor and wall insulation. Austrian households are not highly energy efficient either, and Germany appears to have below-average insulation levels, however these countries are among the richest nations in Europe (using GDP measures) with high levels of social-welfare support and the least afflicted by poverty and income inequality; hence such 'efficiency gaps' are not as pernicious in these countries as in the UK and Ireland where poverty and inequality are far greater (Ramprakash, 1994). As such, the relatively poor thermal-efficiency standards of Ireland and the UK are of more public-health concern than the fair thermal standards found in Germany and Austria because of the differing welfare systems in place in these countries and the consequent lower levels of income poverty and inequality. It could even be argued that the lacklustre energy-efficiency standards of British and Irish housing are of more consequence than the very poor thermal standards in southern-European countries because the interaction in temperature between indoor and outdoor environments may be considerably less harmful in certain Mediterranean nations where seasonal variations in mortality are often less pronounced.

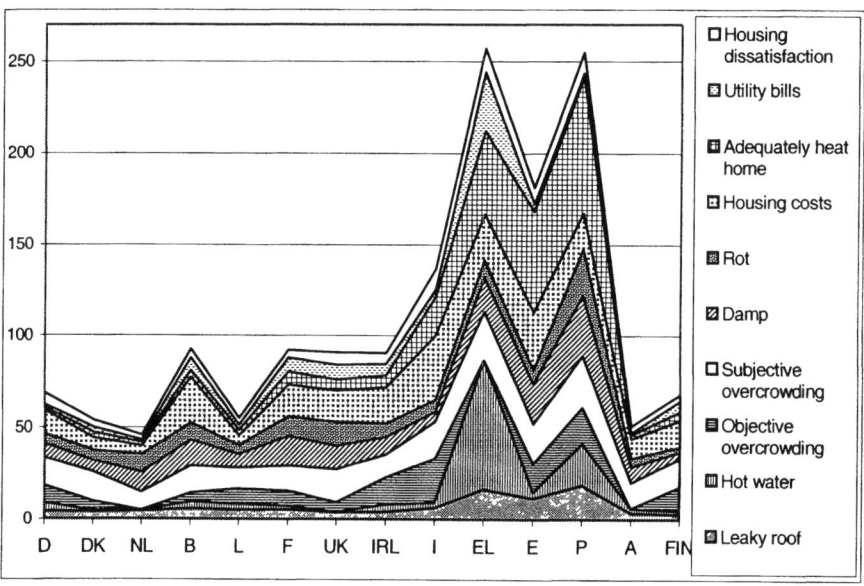

Figure 2.2 Housing conditions: cumulative results (mean %, 1994-97)

A key policy implication of this inefficient use of energy in southern Europe and the UK and Ireland relates to environmental agreements on stabilisation of greenhouse-gas emissions and acidification precursors (the Kyoto and Gothenburg Protocols respectively). Most countries in Europe are required to reduce their energy-related environmental emissions by 2010, with the notable exceptions being the 'Cohesion' countries which have been allocated an increase over 1990 levels. These policy targets are challenging for policy-makers and require dramatic reductions in business-as-usual levels of energy-related environmental emissions across Europe. It can be seen from this research that there is a significant 'efficiency gap' regarding domestic energy-efficiency in Europe, especially in southern Europe, the UK and Ireland. This is of particular interest for Ireland given that its spectacular economic success over the past decade has made its Kyoto and Gothenburg emissions' targets for 2010 very formidable due to the strong link between economic growth and energy use.

A particularly disturbing finding is the level of financial hardship which exists in relation to meeting housing repayments across Europe, and notably in relatively prosperous countries such as Italy, Belgium, France, Ireland and the UK. High levels are also reported in poorer, southern countries like Spain, Greece and Portugal. In addition, state provision of home-heating costs appears to be highest in Finland, Denmark, Spain, Portugal and Germany. Conversely, tenants in Greece and the UK are far more likely to have to meet the costs of their own fuel bills, and these two countries, perhaps unsurprisingly, demonstrate the highest levels of late-payment of utility bills. These findings carry some caveats. Self-perception of financial hardship is very subjective and depends on a variety of factors including household income, the dwelling itself and it's thermal efficiency and the purchasing parity-adjusted price of fuel in a given country.

The findings in this chapter relating to housing satisfaction are also cause for concern. The levels of housing dissatisfaction reported in Greece, Italy and Portugal, in particular, are worrying, for housing is generally regarded as a key indicator for the quality of life of a given population, as well as being fundamental for subsistence. However, as with all of these self-reported data, it is important to be aware of cultural factors which may be influencing the interpretation of the survey questions and, thus, reduce the robustness of cross-country results and comparisons.

Chapter 3

Fuel Poverty in the EU: Developing a Composite Measurement

Introduction

Chapter 2 presented the cross-country results of the study regarding housing deprivation, thermal efficiency, housing affordability and housing satisfaction. This chapter builds on the previous chapter's findings by analysing European data on housing conditions pertaining specifically to fuel poverty.

Fuel poverty, formally identified first in the early 1970s, has continued to exist as a significant social ill three decades on. The extensive debate surrounding fuel poverty in the 1980s and 1990s, where it had been argued that affordable warmth is a "right rather than a privilege", has become poignant once again in the light of unstable and rising worldwide fuel prices. Fuel poverty may be considered different to general (income) poverty. Income poverty can be eradicated efficiently through income support (as has been noted by Boardman, 1991), whereas the eradication of fuel poverty requires not just income subsidisation but also crucial investment in the capital stock (i.e. the household), as fuel poverty is caused by a complex interaction between low income and domestic energy inefficiency. A straightforward *ad hoc* policy of income transfers (such as the fuel allowance in Ireland) may result in more people being able to afford adequate home heating, however it is likely that such funds will be used inefficiently. Energy inefficient households will spend more on fuel to achieve adequate household temperatures rather than invest in improvements to the building fabric which will reap long-term net gains. This is because of market failures which are most notably in the form of information gaps, i.e. households often are unaware of the benefits – or existence – of energy-saving measures, and financial constraints (including access to capital).

When fuel becomes more expensive (as it has done in Europe in particular over the past two to three years), households find it increasingly difficult to heat their homes adequately. Those living in energy inefficient dwellings and on low incomes often cannot afford to heat their homes adequately and suffer from fuel poverty. The problem is a serious political, environmental, social and public-health issue and the UK government, in particular, has invested considerably in assisting low-income groups upgrade their housing. However, although recent estimates show some decline in the levels of fuel poverty in the UK, they confirm the persisting nature and considerable scale of the problem (DEFRA and DTI, 2001). In addition, fuel poverty has many attendant effects, most notably on human health

(Clinch and Healy, 1999b; Rudge and Nicol, 2000).[1] Yet, the net benefits to society (in terms of reduced energy consumption and environmental emissions and improvements in health and comfort) of eradicating fuel poverty through implementing domestic energy-efficiency programmes are very substantial (Clinch and Healy, 2001).

Furthermore, there has been virtually no comparative empirical analysis of fuel poverty undertaken hitherto. This research deficit has occurred, not because of a lack of interest in the area, but because of some major logistical reasons, most obviously the lack of comparable cross-country data. This has now been rectified with the availability of the European Community Household Panel. Data from this extensive, longitudinal survey relating to housing conditions and home heating have been analysed so as to quantify levels of fuel poverty across Europe using a Consensual approach; these data are mainly indicators of housing deprivation which are based, *à la* Townsend, on socially perceived necessities. As such, this study represents the first pan-European quantitative analysis of fuel poverty using comparable, longitudinal data over the years 1994-97 inclusive. A variety of objective and subjective indicators of fuel poverty are utilised and a composite (weighted) measurement is derived. Multivariate Probit regression analysis is conducted for each indicator to validate the cross-tabulation results and to examine those factors that influence the probability of being fuel poor. Sensitivity analysis is also conducted to test the effects of changing various methodological assumptions (including altering the respective weights assigned to each indicator of fuel poverty), and socio-economic and socio-demographic analysis conducted towards the end of the chapter identifies precisely, for the first time, who is suffering from fuel poverty across Europe. First, the context of fuel poverty and energy inefficient housing in Europe is delineated using cross-country data on energy-efficiency standards and energy consumption in European housing.

The European Context

The context for European fuel poverty is important for a number of reasons. From an architectural basis, energy-efficiency standards vary considerably across Europe. The British and Irish housing stocks appear to be among the most energy inefficient in northern Europe when examined using multiple criteria, as has been hypothesised in the past (Boardman, 1991). However, the data presented in Chapter 2 also indicate problems in the Austrian and Belgian stocks, both of which are relatively energy inefficient. From Table 2.11 it can be seen that levels of cavity-wall insulation are considerably below the EU-average in both the UK and Ireland. Ireland has the lowest level of double-glazing in northern Europe, with just one-third of all dwellings fitted with this measure; France, Austria and the UK are also below-par in this regard. The UK has the lowest level of floor insulation in northern Europe, with just 4% of dwellings equipped. The only energy-efficiency

---

[1] Although it is important to note that the research investigating the health effects of fuel poverty is slow to identify causality, and is far from conclusive.

measure with which Irish and British households perform satisfactorily is roof insulation. Conversely, countries such as Denmark, Finland, Sweden and Norway demonstrate exemplary thermal-efficiency standards.

Of course, Scandinavian countries prioritise these measures in the design and construction of new housing, as it is essential protection to combat the relatively severe winters experienced in these colder climates where winter temperatures are often below freezing (Clinch and Healy, 1999a). Nonetheless, Ireland and the UK have the highest rates of seasonal mortality in northern Europe, and it has been shown that such mortality rates result, in no small part, from the inadequately protected, thermally inefficient housing stocks in these countries (Clinch and Healy, 2000c; Eurowinter Group, 1997). There are also strong associations between inadequately heated homes and increased rates of morbidity; higher incidences of various cardiovascular and respiratory disorders have been associated with cold exposure from within the home (Collins, 1986 and Evans *et al.*, 2000). Thus, when temperatures fall during a typical British or Irish winter, households need to increase their expenditure on fuel considerably to heat their home adequately, owing to the poor level of heat retention in their homes. The problem of fuel poverty occurs, therefore, when a low-income household lacks adequate insulation levels and an efficient heating system to achieve affordable warmth.

Data on energy-efficiency standards in southern Europe are notoriously difficult to obtain, and their reliability is often questionable. Eurostat have provided robust data for Greece, but only scant data for Portugal and no data whatsoever for Italy or Spain. However, the data that is available, together with the analysis of general housing conditions conducted by Healy (2001a), indicate that energy efficiency in southern-European housing is not a priority in their building regulations. It is an often-overlooked fact that many parts of southern Europe also face cold winter temperatures,[2] yet their housing stocks appear to be highly energy inefficient, and they are also the poorest countries in Europe (using measures such as income poverty and inequality as well as macroeconomic indicators like GDP per capita).[3] Despite this, there has been virtually no published research on fuel poverty in southern Europe. This chapter attempts to rectify this situation.

Many countries demonstrating poor levels of domestic energy efficiency are consuming greater amounts of energy than necessary, as people inhabiting inefficient dwellings must consume more fuel to heat their homes. Irish energy consumption per household is the highest in northern Europe, at 102,000 Megajoules per annum, compared to an EU-13 average of 77,000 MJ. Consequently, environmental emissions, such as CO_2, SO_2, NO_x and particulate matter are also greater. This is of considerable importance given that many European countries – including Ireland – are having extreme difficulty in meeting their agreed targets for stabilisation of greenhouse-gas emissions under the Luxembourg Agreement and acidification precursors under the Gothenburg Agreement (Clinch and Healy, 2000b).

---

[2] Data on this is presented in Chapter 9 (see Table 9.3).
[3] Italy being the exception here, with GDP per capita considerably above the EU-average.

Measuring Fuel Poverty

Fuel poverty can be measured in a number of ways and three major approaches are now outlined.

Temperature

The earliest research on fuel poverty based its discussion on "adequate home heating" and "adequate warmth" (e.g. Lewis, 1982 and Boardman, 1986). There are a range of temperatures that may be defined as those within an adequate heating regime. Using a definition based on temperature, fuel poverty may be calculated by quantifying those households which fail to achieve minimum 'adequate' levels of household warmth. Such an approach, though simple to conduct in theory, is problematic for a number of logistical and scientific reasons, chiefly because of the inadequacy and unreliability of data on household temperatures, while intermittent occupancy may also distort results using this approach.

Expenditure

This approach was first advocated by Boardman (1991). A simple 'fuel-poverty line' can be set – similar to that in poverty research – where households are considered fuel-poor if they spend more than X% of their income on home heating. Generally, a 10% threshold based on net household income has been used; this definition has recently been changed so that the denominator excludes housing costs. This has been considered a standard definition of fuel poverty and has been employed by the UK government in much of its analysis on the subject (DETR, 1999). However, such an approach has a number of flaws, especially regarding the non-existent scientific rational behind setting the budget line at 10% of net income; there has been no attempt to justify this threshold as appropriate and robust and there is little published theoretical debate on such matters. The approach is also rendered ineffective in cross-country analysis of fuel poverty where differing levels of purchasing power and differing real and nominal fuel prices would reduce the comparability of such a measurement.

Consensual

This approach follows the method pioneered by leading poverty and deprivation researchers, Peter Townsend and David Gordon (Townsend, 1979, Gordon et al., 2000). A basket of 'goods' that are regarded as socially perceived necessities may be employed as a suite of indicators to calculate fuel poverty. Some of these necessities fall under the umbrella of fuel poverty; the absence of certain items regarded as essential household attributes (or socially perceived necessities) may be considered indicators of fuel poverty, as can the presence of certain unwanted household characteristics. For example, possession of central heating in the home is considered by the vast majority of the population in the UK and Ireland to be a necessity, as is a damp-free home (Callan *et al.*, 1993, Gordon *et al.*, 2000). The lack of central heating and the presence of damp act as indicators of fuel poverty using an approach founded on consensual social indicators.

Such an approach attempts to capture the wider elements of fuel poverty, such as social exclusion and material deprivation, as opposed to approaches based solely on home-heating expenditure or household temperature. It also benefits over the standard income-based approach in that it is based on households' actual feelings and statements (i.e. it is valuable in eliciting experiential fuel poverty), as opposed to being based solely on an arbitrary calculations or estimations attempting to identify those who might be fuel-poor. It can also be adapted over time to reflect changes in attitudes and socially perceived necessities, as well as rising living standards. In this regard, the methodology can be said to be truly dynamic as it can be altered over time to incorporate changes in social norms and expectations. However, it should be noted that there are potential limitations associated with this approach. Perhaps the strongest relates to the margin of error associated with reporting households'/individuals' declarations and assessments regarding personal characteristics. Such error is likely to apply to both subjective and objective indicators of fuel poverty, but especially the former.

While the UK government has traditionally used the standard definition of fuel poverty in measuring levels across the UK, an alternative measurement of fuel poverty is now being evaluated using this Consensual approach to calculate the incidence of fuel poverty (DEFRA and DTI, 2001). Such an approach is developed and employed in this chapter for the UK and 13 other EU countries.

Fuel Poverty in Europe: Cross-Country Results

Data: European Community Household Panel[4]

The European Household Panel (ECHP) is a standardised, multi-purpose and longitudinal survey, providing comparable information across the Member States on income, work and employment, poverty and social exclusion, housing, health and other diverse social indicators regarding the living conditions of private households and persons. The crucial feature of the ECHP is the harmonisation of its methodology and results, through the creation of a centralised questionnaire. During the first Wave (year) of the questionnaire, the collection of data occurred in 12 countries in Europe (all EU Member States in 1994); this increased to 13 in 1995 (when Austria joined), 14 in 1996 (with the inclusion of Finland) and 15 in 1997 (with the participation of Sweden). Consequently, some flexibility was granted to each participating country to adapt common procedures to relate to their own local situations. Data are collected by National Data Collection Units in each country. These Units – normally research institutes or national statistics' centres – tailor the questionnaire to make it suitable for their own respective countries. High response rates (more than 70%) were obtained for all four waves of the study, and

---

[4] Some of this section is taken from Eurostat (1996).

some 60,000 households and 130,000 adults are interviewed successfully in each wave. An outline of the sample size and allocations per country is provided in Table 3.1.

Table 3.1 Sample size and response rate for ECHP in 1994

Country	No of households ('000)	Selected sample	Completed sample	% of households surveyed	Response rate (%)
B	4,100	4,886	4,192	1.0	85.8
DK	2,400	5,500	3,482	1.5	63.3
D	36,900	10,572	5,054	0.1	47.8
EL	4,000	6,131	5,523	1.4	90.1
E	13,700	8,000	7,206	0.5	90.1
F	27,700	11,117	7,344	0.3	66.1
IRL	1,100	7,252	4,048	3.7	55.8
I	12,900	7,989	7,115	0.6	89.1
L	150	2,826	1,011	6.7	35.8
NL	6,500	5,926	5,187	0.8	87.5
P	3,600	6,238	4,881	1.4	78.2
UK	23,800	8,104	5,779	0.2	71.3
EU-12	13,685,000	84,541	60,822	0.4	71.9

Source: Columns 2, 3 and 4 taken from Eurostat (1996); columns 5 and 6 are the author's calculations.

The probability-based (random) sample size for each Member State was determined on the basis of a number of theoretical and practical considerations. Generally, countries with larger populations received larger sample sizes to allow for maximum disaggregation of results. In Germany, legal regulations restricted the sample size to about 5,000 households; as such, Germany has the fewest households sampled as a percentage of the total population in the study. The highest response rate was found in Greece and Spain, where over 90% of households successfully completed the ECHP questionnaire in 1994. The lowest response rate was in Luxembourg, with just one-in-three households agreeing to undertake the interview.

The sample was normally distributed proportionately across geographical regions across each Member State. This enabled maximum precision of estimates at the national level. Italy and Spain, however, chose to sample disproportionately high allocations in smaller regions with a view to ensuring a minimum sample size for each region of the country. Apart from this variation in sampling rates across regions in some larger countries, in general all of the population were sampled at the same rate within each country. Thus, there was no over-sampling of any particular groups. In some cases, certain small parts of the population were not covered because of the sampling frame selected. For instance, in Ireland households were selected from the electoral register which excludes households recently arrived in Ireland and those not covered for some other reason. Households comprised exclusively of persons aged under 18 years are also not covered. In Germany, the Netherlands and Greece, non-nationals unable to speak the national language were also excluded. Collective dwellings were also omitted from the sample frame in Greece. Rural communities of less than 5,000 accommodations were not sampled in France. Foreigners not registered as permanent residents in Italy were excluded. Institutional housing was not sampled in the Netherlands. Finally, new-built Portuguese homes were not sampled.

Each household in the sample is weighted in inverse proportion to the probability with which it was selected. The design weights are incorporated into the methodology to compensate for differences in the probabilities of selection into the sample, with the weight given to each household being inversely related to its probability of selection. With multi-stage sampling design employed, it was important that the reference is to the overall (relative) selection probabilities of households, not the absolute values. In application, weights are scaled such that the mean value per household (with interview completed) is 1.0. If p_i is the overall sampling probability of household 'i' and n_i the number of households successfully enumerated in the sample, then the design weights can be denoted as follows:

$$w_i = \frac{1}{p_i} \left(\frac{\sum n_i}{\sum n_i / p_i} \right)$$

The weights are calculated for all households selected, but the summation and n_i on the right is confined to households with the interview completed successfully. This ensures that the weights are normalised, i.e.

$$\Sigma w_i . n_i = \Sigma n_i$$

Non-response rates are introduced to reduce the effect of differences in unit response rates in different parts of the sample. These weights are based on known characteristics of the sample. Weighting for non-response is particularly important when rates of non-response are high and generally variable from one part of the population to another. It was found that ECHP non-response rates are often high with large variations across population groups. To correct for gross distortions in the distribution of the achieved sample, it was necessary to weight for non-response according to various characteristics of the households. Weighting involves the division of the sample into certain appropriate weighting classes, and within each weighting class the weighting-up of the responding units in inverse proportion to the response rate so as to compensate for the non-responding cases in that class.

The non-response weight (for a single classification) is computed as follows:

$$w_j = \frac{\bar{R}}{R_j}$$

The numerator is the overall response rate for the household interview (i.e. n_j as the number of interviewed cases in class j:

$$\bar{R} = \frac{\sum n_j}{\sum \left(n_{j/R_j}\right)}$$

Weights are applied also to correct distributional problems with households and persons in the sample frame. Extreme values are trimmed to avoid inflation in variances.

The data used in this study come from all currently available (four) Waves of the ECHP, undertaken in 1993-94, 1994-95, 1995-96 and 1996-97. However, data on successive years will become available over time. This data was purchased directly from Eurostat Luxembourg using a tailored research contract for the Department of Environmental Studies, University College Dublin. More details of the survey and its methodology can be found in Eurostat (1996).

Methodology

The methodology represents a very different approach to that traditionally used. In this study fuel poverty is quantified for 14 European countries using six social indicators of fuel poverty, all of which are asked in the ECHP. The indicators pertain to household finances (fuel and utility bills), the building fabric (presence of damp, rot, etc.) and the heating system fitted in the house. The six indicators are split into two sub-groups: subjective (self-reported) indicators based on householders' declarations and feelings about their housing and income, and objective indicators based on factual characteristics or conditions of the dwelling and their finances which are not skewed by potential subjective biases.

Although it would have been useful to include an indicator on access to energy markets – this is becoming more and more important in increasingly liberalised European energy markets – such an indicator is not available in the ECHP. It is suggested that future Eurostat surveys incorporate such an indicator so that the dynamic definition of fuel poverty can be continuously updated to reflect changing needs and societal expectations; this is actually one of the strongest features of the Consensual approach to poverty estimation. The composite measurement developed here encompasses a range of indicators that are central to the fuel poverty problem.

With this in mind, the results of each fuel-poverty indicator is discussed in detail now, starting with the subjective indicators. Probit multivariate regression models are presented later.

Subjective Indicators

Households Unable to Afford to Heat Home Adequately Fuel poverty can be defined as "the inability to afford adequate warmth in the home" (Lewis, 1982). Thus, this indicator is crucially important in estimating levels of fuel poverty, as it encompasses the standard qualitative definition of fuel-poor households. It is, therefore, the key indicator of fuel poverty and is given precedence in the composite measurement of fuel poverty, derived later. Table 3.2 illustrates the results of this key indicator of fuel poverty.

Table 3.2 Households unable to heat home adequately (% of households)

	1994	1995	1996	1997	Mean
Germany	2.0	1.5	1.4	-	1.6
Denmark	4.2	2.9	2.8	2.6	3.1
Netherlands	2.0	1.8	2.0	2.2	2.0
Belgium	4.6	4.1	2.8	3.0	3.6
Luxembourg	2.6	3.1	3.5	-	3.1
France	8.5	7.3	7.0	5.8	7.2
UK	8.9	6.2	5.3	2.7	5.8
Ireland	8.0	5.9	6.5	5.1	6.4
Italy	22.4	22.7	20.6	20.3	21.5
Greece	46.8	45.5	46.8	42.9	45.5
Spain	58.7	57.7	53.3	49.7	54.9
Portugal	75.8	74.9	73.8	72.9	74.4
Austria	-	2.5	1.9	1.8	2.1
Finland	-	-	4.7	4.7	4.7
EU-14	20.4	18.2	16.6	17.8	16.9
EU-10	5.1	3.9	3.8	3.5	4.0

Results show that an alarming 45.5% of households in Greece, 54.9% in Spain and 74.4% in Portugal declare this inability. The variation of results across countries is dramatic: households in Germany, the Netherlands, Austria, Luxembourg and Denmark have no problems heating their homes (all report incidences of 3% or less respectively), while relatively rich nations like Italy and France report substantial levels of this indicator of fuel poverty (21.5% and 7.2% respectively). The average rate throughout all 14 EU countries analysed is 16.9%. A substantially different mean incidence of fuel poverty is found using this indicator across northern Europe: just 4.0%. In this instance, countries such as the UK (5.8%) and Ireland (6.4%) are identifiable as relatively high sufferers of this indicator of fuel poverty. Longitudinally, it is worthwhile to note that, while many countries are realising significant reductions in fuel poverty across the time series, the most notable aggregate decreases in the incidence of fuel poverty are by the UK (a reduction of 70%), Denmark (38%) and Ireland (36%). The success in the UK and Denmark may be largely attributed to the relatively aggressive State-funded fuel-poverty alleviation programmes in place, while in Ireland income growth and rising living standards are more likely to be the root cause.

Households Unable to Pay Utility Bills A household which has been unable to pay on time a scheduled utility bill (gas or electric) over the previous 12 months is likely to be finding it difficult to keep the home adequately heated and, as such, this indicates the potential existence of fuel poverty. Those unable to keep up to date with utility bills may also have suffered disconnection from the supplier, compounding the experience of fuel poverty. The results demonstrate that Mediterranean nations do not appear to suffer as highly with this indicator, even though they report the highest incidences of being unable to heat their homes adequately. Indeed, Italy, Spain and Portugal rank among the best countries in this exercise, however this indicator exhibits the lowest incidence across all countries in the study.

Table 3.3 Households unable to pay scheduled utility bills (% of households)

	1994	1995	1996	1997	Mean
Germany	2.2	1.6	1.5	-	1.8
Denmark	3.2	2.6	2.4	2.6	2.7
Netherlands	1.4	1.1	1.2	1.4	1.3
Belgium	7.7	6.1	6.9	6.7	6.9
Luxembourg	3.1	2.2	2.8		2.7
France	8.8	7.2	7.3	7.0	7.6
UK	9.4	7.8	7.0	-	8.1
Ireland	8.4	6.3	6.1	4.9	6.4
Italy	3.8	3.8	4.5	4.1	4.1
Greece	36.5	30.1	-	29.3	32.0
Spain	5.2	4.3	3.7	4.1	4.3
Portugal	3.1	1.7	1.7	1.8	2.1
Austria	-	1.1	1.1	1.2	1.1
Finland	-	-	11.4	0	5.7
EU-14	7.7	5.8	4.4	5.7	6.2
EU-10	5.5	4.0	4.8	3.4	4.4

The highest level of late-payment of utility bills is found in Greece, where a remarkable 32% are affected. The UK (8.1%) and France (7.6%) also perform poorly. The mean (EU-14) proportion of households unable to heat their homes on time is 6.2%, while the corresponding statistic for the northern Europe is 4.4% (Table 3.3). The time-series shows that the UK has reduced the proportion of households unable to pay utility bills on time by a quarter, while Ireland has achieved an even larger reduction of 42%. Spain also reports positive longitudinal results: a reduction of 21% over four years. Other countries report less clear results, while some are showing increased difficulties in meeting the costs of utility bills.

Households Lacking Adequate Heating Facilities Household declaring inadequate heating systems cannot, by definition, heat their home satisfactorily or efficiently; thus, this is a good subjective experiential indicator of fuel poverty. The results show that households declaring inadequate heating facilities are found in high numbers in certain southern countries (Greece and Portugal) but also in wealthier, colder, northern nations like France and the UK. An overall (EU-14) incidence of 11.5% is calculated for this indicator. The highest incidence of inadequate heating facilities is found among Portuguese households, where two-in-five such households (39.7%) declare this indicator, while Greece (34.3%) and Italy (17.9%) also fare badly. In northern Europe a rate of 6.7% is reported across all households, on average, while France performs poorest in this set of nations (one-in-nine households), with the UK (9.7%) a close second, though data fro the latest available year (1997) show a marked decrease. The longitudinal pattern across Europe shows some substantial decreases in the incidence of this indicator, but also some increases (e.g. Portugal). The biggest downward trends are found in the UK (a 62% reduction between 1994 and 1997) and France (a 15% fall during the same period).

Table 3.4 Households declaring a lack of adequate heating facilities (% of households)

	1994	1995	1996	1997	Mean
Germany	6.1	4.6	3.7	-	4.8
Denmark	4.0	3.5	4.1	3.4	3.8
Netherlands	6.6	5.9	6.9	6.0	6.4
Belgium	8.7	7.3	8.1	7.6	7.9
Luxembourg	5.2	5.2	5.6		5.3
France	12.4	10.8	10.3	10.5	11.0
UK	14.3	10.0	9.1	5.5	9.7
Ireland	9.6	7.4	7.6	7.0	7.9
Italy	21.1	17.7	16.1	16.5	17.9
Greece	39.1	36.0	30.8	31.3	34.3
Spain	5.0	1.2	1.3	2.5	2.5
Portugal	39.2	39.3	40.1	40.3	39.7
Austria	-	7.2	6.7	4.8	6.2
Finland	-	-	3.0	2.8	2.9
EU-14	14.3	12.0	11.0	11.5	11.5
EU-10	8.4	6.9	6.5	6.0	6.6

This indicator acts as an interesting comparative cross-check with the objective indicator of fuel poverty, 'Households Lacking Central Heating' (Table 3.6). The disparity demonstrates the lack of awareness among householders regarding what it is that represents an 'adequate' heating system. If householders were fully informed, only centrally heated and electric-storage heating systems would be considered efficient, adequate means of home heating. Nonetheless, despite some degree of mismatch between the subjective and objective indicators, this indicator is considered useful in eliciting experiential fuel poverty. The results, reported in Table 3.4, show that many households believe their system is adequate when, on an economic-efficiency basis, they are not. The widest disparities are found in Portugal, Spain and Greece.[5]

Objective Indicators

Presence of Damp Walls and/or Floors The presence of damp indicates that the dwelling is not energy efficient. It may also be a manifestation of a continuously unheated or ineffectively heated home. In either case, it acts as a good objective indicator of fuel poverty. Some 12.7% of all European households contain damp patches. Again, Greece (18.8%), Spain (21.5%) and Portugal (33.4%) are suffering worst in this regard. In northern Europe an average incidence of 9.8% is found, and the UK (13%), Belgium (14.4%) and France (16.3%) all perform poorly with regard to the presence of damp.

These results, tabulated in Table 3.5, are particularly important from a public-health perspective, as chronic exposure to damp is strongly associated with ill-health, mainly via increased respiratory and cardiovascular diseases (Williamson *et al.*, 1997). There have been some dramatic reductions in the presence of damp throughout the 1990s. The UK has experienced a 53% fall in damp spores, while the level of household damp has fallen by 44% in Italy and by 28% in France between 1994 and 1997. A potential caveat with this indicator relates to the fact that damp can be caused by inadequate ventilation as well as inadequate heating. It is worth bearing this in mind when considering the findings reported in this section.

---

[5] More on this disparity can be found in the proceeding section on 'objective' indicators.

Table 3.5 Households with damp spores (% of households)

	1994	1995	1996	1997	Mean
Germany	9.5	7.8	6.4	-	7.9
Denmark	6.6	6.2	6.5	5.6	6.2
Netherlands	12.0	12.0	9.8	9.5	10.8
Belgium	15.8	16.7	12.3	12.8	14.4
Luxembourg	7.2	8.2	7.2	-	7.5
France	19.5	17.1	14.6	14.0	16.3
UK	17.2	14.3	12.2	8.1	13.0
Ireland	10.5	9.4	8.9	9.4	9.6
Italy	7.2	5.4	4.8	4.1	5.4
Greece	20.8	17.7	18.5	18.2	18.8
Spain	25.6	19.2	20.4	20.8	21.5
Portugal	32.7	32.3	33.5	35.2	33.4
Austria	-	10.1	8.3	8.1	8.8
Finland	-	-	3.9	3.7	3.8
EU-14	15.4	13.6	12.0	12.5	12.7
EU-10	12.3	11.3	9.0	8.9	9.8

Lacking Central Heating Households not possessing central heating or similar heating systems generally find it more difficult to efficiently heat the home. It was shown by Clinch and Healy (1999a) that other heating systems, such as solid fuel and LPG, are dearer, dirtier and less efficient, and are generally possessed by low-income households. As such, they act as economically regressive generators of home heating. The lack of either central heating or electric-storage heating is a good objective indicator of fuel poverty. It is interesting to note that this indicator of fuel poverty is reported at a repeatedly higher incidence than any other indicator across all countries in the study, yet the provision of central heating is one of the most effective measures in eradicating fuel poverty.

Some 91.7% of Portuguese households do not have central heating fitted in their homes, whilst the corresponding percentages for Spain and Greece are 67.2% and 45.7%.[6] In northern Europe the highest incidence of this fuel-poverty indicator are located in Belgium (23.9%) and Ireland (20.2%). An average rate in this group of ten countries of 12.5% is calculated, while across all Europe one-in-four households (25.0%) fail to possess central heating or electric storage heaters (Table 3.6). Across Europe, the penetration of central heating is increasing almost universally, but not at a similar rate of penetration. Countries formerly lacking in such measures are catching up; for instance, Ireland has increased its penetration of central heating by 31% over the four-year period, 1994-97, and a similar rate is found for the UK (32%). Elsewhere in northern Europe, Austria has reduced its proportion of households without central heating by almost a half (49%) during the

---

[6]Yet, southern-European countries suffer from cold winters which, though not as severe as in northern Europe, are still cold enough for fuel poverty and cold-related ill-health (Healy, 2003).

four years, 1994-97. In southern Europe Spain has progressed, with a 13% reduction in households lacking central heating, though other southern countries have not made such advances.

These results are interesting when they are cross-checked with the corresponding subjective indicator of adequate heating facilities. The disparity between what the public believes constitutes an adequate heating system and what actually is adequate in energy-efficiency literature is interesting for another reason: it demonstrates one key cause of market failure – the 'information gap' – in the market for domestic energy-efficiency measures.[7] The exceptions to the trend are Denmark, Finland and France, where households over-declare inadequate heating facilities.

Table 3.6 Households without central or electric-storage heating (% of households)

	1994	1995	1996	1997	Mean
Germany	12.4	10.3	8.6	15.3	11.7
Denmark	3.2	1.8	1.0	1.0	1.8
Netherlands	14.4	13.2	12.7	11.4	12.9
Belgium	25.8	24.3	24.0	21.4	23.9
Luxembourg	7.6	6.9	5.4	-	6.6
France	10.6	10.1	9.8	8.9	9.9
UK	14.8	12.1	10.7	10.0	11.9
Ireland	23.8	20.8	19.8	16.4	20.2
Italy	26.6	18.6	17.3	17.7	20.1
Greece	44.7	46.9	46.4	44.7	45.7
Spain	71.6	69.9	65.2	62.2	67.2
Portugal	92.5	92.1	92.2	89.9	91.7
Austria	-	19.8	17.9	10.0	15.9
Finland	-	-	3.4	16.3	9.9
EU-14	29.0	26.7	23.9	25.0	25.0
EU-10	14.1	13.3	11.3	12.3	12.5

Rotten Window Frames Window frames which have become rotten are not energy efficient and can be considered a good (objective) indicator of fuel poverty. Householders are asked whether some or all of their windows contain significant signs of rot. Rot is most commonly found in Portugal where a quarter (25.2%) of all households have rotten windows, compared with an EU-14 mean incidence of 8.6%. In northern Europe 12.7% of British households are suffering from this indicator of fuel poverty, and in France the corresponding percentage is 10.4% (Table 3.7). This compares to a northern-European mean of 7.2%. There have been very substantial decreases in the proportion of households with rotten windows

---

[7] Clinch and Healy (2000b) outline the wide range of reasons for this market failure, and the information gap – where households are not aware of the full benefits of energy-efficiency measures – is considered a major explanation for this phenomenon.

over the four-year period. The reduction in the UK represents a 35% fall on 1994 levels, while in Ireland the corresponding fall is 25%. In southern Europe Greece has reduced its level of rotten windows by 37%.

Table 3.7 Households with rotten window frames (% of households)

	1994	1995	1996	1997	Mean
Germany	6.4	5.4	4.2	-	5.3
Denmark	6.1	5.2	5.8	5.2	5.6
Netherlands	9.8	10.2	9.8	8.8	9.7
Belgium	9.2	10.3	8.7	8.3	9.1
Luxembourg	4.8	5.2	4.4	-	4.8
France	11.2	10.7	9.7	10.0	10.4
UK	15.3	14.1	11.6	9.9	12.7
Ireland	8.9	6.4	7.0	6.7	7.3
Italy	8.0	6.2	5.2	5.3	6.2
Greece	11.8	9.6	8.5	7.4	9.3
Spain	9.7	7.7	6.4	6.6	7.6
Portugal	24.2	26.1	25.0	25.3	25.2
Austria	-	5.5	4.4	3.8	4.6
Finland	-	-	2.5	2.6	2.6
EU-14	10.5	9.4	8.1	8.3	8.6
EU-10	9.0	8.1	6.8	6.9	7.2

Composite Measurement of Fuel Poverty

A variety of aggregate measurements of fuel poverty have been derived. These results are weighted (composite) estimates of fuel poverty. Each indicator is assigned a weight, and each weight varies in the sensitivity analysis in accordance with their relevance to the qualitative definition of fuel poverty. Thus, 'Inability to Afford Adequate Home Heating', as the key indicator of fuel poverty, may be given a higher weighting than the other indicators, and so forth. A variety of sensitivity analyses are conducted to test the sensitivity of the results to various assignments of weights. For economies of space and ease of reading, the indicators are denoted in the weighting equations as follows:

α = Unable to afford to heat home adequately
β = Unable to pay utility bills on time
π = Lack of adequate heating facilities
δ = Damp walls and/or floors
λ = Rotten window frames
μ = Lacking central heating

Of course it should be noted that certain indicators are highly correlated. π is a subjective indicator which is highly associated with μ; α and β are similarly correlated, and δ and λ are likely to be strongly related to each other. Weighting scenarios were developed following consultations with practitioners in the area of fuel poverty. A steering committee encompassing practitioners, academics and policymakers was set up in order to develop scenarios where weights were plausible and meaningful.

Scenario 1: Key Indicator Given Strong Preference Here, the key indicator of fuel poverty, 'Households unable to afford to heat their home adequately' (α), is given a weight of 0.5; each of the five subsequent indicators is assigned a weight of 0.1 respectively, i.e.

$$0.5\,\alpha + 0.1\,\beta + 0.1\,\pi + 0.1\,\delta + 0.1\,\lambda + 0.1\,\mu$$

This scenario is thought to be especially useful for comparisons across Europe, as indicators relating to damp and rotten windows may not be as pertinent in southern-European countries Under this analysis, the highest incidences of composite fuel poverty in southern Europe include: Portugal (56.4%), Spain (37.8%), Greece (36%) and Italy (16.1%). In northern Europe, France (9.1%) the UK (8.4%) and Ireland (8.3%) appear to have the highest rates of composite fuel poverty. The mean rate of fuel poverty across EU-14 is 14.8%, and in northern Europe a mean of 6% is found.

Scenario 2: Equal Weights If it is thought that all indicators are of equal importance, then equal weights can be assigned to each of the six indicators. Under this scenario, all indicators are given a weighting of 0.17 respectively, i.e.

$$0.17\,\alpha + 0.17\,\beta + 0.17\,\pi + 0.17\,\delta + 0.17\,\lambda + 0.17\,\mu$$

Under this scenario, rates of composite fuel poverty in southern Europe are decidedly lower: Portugal (44.4%), Greece (29.7%), Spain (26.3%) and Italy (12.5%). In northern Europe relatively high rates are calculated for Belgium (11%), France (10.4%), the UK (10.2%) and Ireland (9.6%). The average EU-14 rate is calculated as 13.4%, while the northern-European rate is found to be 7.4%.

Scenario 3: Subjective Indicators Only It may be useful to consider the subjective and objective indicators separately. Disaggregating in this manner, so that only subjective social indicators of fuel poverty are considered, implies giving a weighting of 0.33 to each of the three subjective indicators, i.e.

$$0.33\,\alpha + 0.33\,\beta + 0.33\,\pi$$

The four southern-European countries once more demonstrate the highest composite levels of fuel poverty. Under this scenario, Portugal has a rate of 38.7%, followed by Greece (34.7%); both of these levels are lower than the previous two scenarios. However, Spain (20.6%) and Italy (14.5%) display slightly higher levels of composite fuel poverty under this scenario. In northern Europe the highest rates are found in France (8.6%), the UK (7.9%) and Ireland (6.9%). An EU-14 average of 11.3% is calculated, while in northern Europe the mean is found to be 5%. These results, being the lowest found in the sensitivity analysis, act as lower bound estimates of fuel poverty.

Scenario 4: Objective Indicators Only Here, the indicators are disaggregated in the same manner as in Scenario 3, but in this case only objective social indicators of fuel poverty are analysed. Again, an equal weighting of 0.33 is assigned to each indicator, i.e.

$$0.33\,\delta + 0.33\,\lambda + 0.33\,\mu$$

These results fall in the upper bound tail of estimates of fuel poverty for northern Europe, but not for southern countries. High levels are reported in Portugal (50.1%), Spain (32.1%) and Greece (24.6%). In northern Europe, the highest levels are found in Belgium (15.8%), the UK (12.5%) and Ireland (12.4%). A relatively low EU-14 average of 15.4% is calculated, together with a relatively high EU-10 average (for northern Europe) of 9.8%.

Scenario 5: Key Indicator and Objective Indicators Given Preference If it is felt that the key indicator and the objective indicators are more reliable than the subjective indicators, then weights may be distributed as follows:

$$0.5\,\alpha + 0.17\,\delta + 0.17\,\lambda + 0.17\,\mu$$

This leads to the very highest results in southern Europe, with composite levels of fuel poverty of 62.8% in Portugal, 43.8% in Spain, 35.3% in Greece and 16.1% in Italy. Rates of 9.9% in Belgium, 9.8% in France and 9.5% in Ireland are also relatively high. An average rate of 16.3% of all households in EU-14 is calculated, while in northern Europe a rate of 7.0% is found under this scenario.

Scenario 6: Key Indicator Given Moderate Preference While it seems wise to weight the key indicator of fuel poverty higher than other indicators, a weight of 0.5 may seem too generous. Hence, in this case, the first scenario is altered so that a weight of one-third is given to the key indicator and all other five indicators are assigned equal weights of 0.134, i.e.

$$0.33\,\alpha + 0.134\,\beta + 0.134\,\pi + 0.134\,\delta + 0.134\,\lambda + 0.134\,\mu$$

Table 3.8 Sensitivity analysis of fuel poverty results (% of households, 1994-97)

Scenario	1	2	3	4	5	6
Germany	4.0	5.5	2.7	8.3	5.0	4.7
Denmark	3.6	3.9	3.2	4.5	3.9	3.7
Netherlands	5.1	7.2	3.2	11.1	6.7	6.2
Belgium	8.0	11.0	6.1	15.8	9.9	9.5
Luxembourg	4.2	5.0	3.7	6.3	4.8	4.6
France	9.1	10.4	8.6	12.2	9.8	9.8
UK	8.4	10.2	7.9	12.5	9.3	9.3
Ireland	8.3	9.6	6.9	12.4	9.5	9.0
Italy	16.1	12.5	14.5	10.6	16.1	14.3
Greece	36.0	29.7	34.7	24.6	35.3	32.8
Spain	37.8	26.3	20.6	32.1	43.8	31.9
Portugal	56.4	44.4	38.7	50.1	62.8	50.3
Austria	4.7	6.5	3.1	9.8	6.0	5.6
Finland	4.8	4.9	4.4	5.4	5.1	4.9
EU-14	14.8	13.4	11.3	15.4	16.3	14.0
EU-10	6.0	7.4	5.0	9.8	7.0	6.7

Under this final scenario, middle bound estimates are derived for both northern and southern Europe. Portugal reports a composite rate of 50.3%, while the respective rates in Greece (32.8%), Spain (31.9%) and Italy (14.3%) demonstrate a centralised estimate of the range produced under the sensitivity analysis. In northern Europe France is calculated to have the highest level of fuel poverty using this scenario, with 9.8% affected. However, Belgium (9.5%), the UK (9.3%) and Ireland (9.0%) are all showing very similar rates. The average incidence of fuel poverty in northern Europe is 6.7%, while the corresponding incidence for all 14 countries in this analysis is 14.0%. These results appear to be robust and are used as good middle-bound estimates of fuel poverty (Table 3.8).

Socio-Demographic and Socio-Economic Analysis

A socio-demographic and socio-economic profile of those households unable to afford adequate heat in their home – the key indicator of fuel poverty – now follows. This isolates those groups in society most vulnerable and at risk of suffering fuel poverty.

Sociological Type

Some households need more fuel than others because their circumstances dictate that the home must be heated for longer intervals or because they require higher temperatures (e.g. households occupied by the elderly or those with very young children). For some households, heating costs may be disproportionately high because these costs may fall on one person (e.g. single-adult households). Furthermore, there are some households that are known – through poverty and deprivation research – to face severe financial hardship (e.g. single-parent households). All of these factors affect the probability of certain household groups enduring fuel poverty. The socio-economic and socio-demographic analysis in this section examines a variety of household types based on demographic (e.g. marital status, sociological type, house type) and economic (e.g. main income source, housing tenure) variables.

For economy of space, the mean results (over the four-year time series) are presented henceforth. Table 3.9 presents the incidence of fuel poverty across Europe by social group using an amalgamation of Eurostat's taxonomies for Sociological and Economic typologies; this allows for maximum disaggregation of the results across EU-14 whilst retaining statistical robustness.

Lone Parents The results show that, across Europe, the group most at risk of suffering from persistent fuel poverty is single parents, especially those whose children are all under 16 years of age, where an average (EU-14) incidence of 21.8% is found. In northern Europe an average (EU-10) incidence of fuel poverty of 10.3% is calculated for this household group. The highest incidence is calculated in Portugal, where some 71.4% of single parents are declaring fuel poverty. Spain fares particularly poorly also, with 62.7% of this category of single parents affected. The proportion in Greece is also high, at 48.5%. The Irish rate of fuel poverty for single parents (for all children under 16) is 19.3%, making this the highest level in northern Europe. Elsewhere in northern Europe, single parents exhibit high levels of fuel poverty in the UK (18.6%) and France (15%). Single parents with at least one child over 16 generally demonstrate lower levels of fuel poverty. This is probably because of their improving financial situation as their children grow up and become less dependent. However, the results demonstrate only slightly reduced incidences among this household group. The problem of fuel poverty among lone-parent households is acute for two causal reasons: first, they generally suffer from low incomes and will therefore find it difficult to make ends meet regarding fuel bills; second, their financial circumstances entail that they are more likely to live in poor (energy inefficient) housing which makes home heating even less affordable.

Lone Pensioners Both male and female pensioners exhibit high levels of fuel poverty. However, there is a gender bias towards females, with 22.1% of female lone pensioners suffering from fuel poverty across Europe, compared with 20.4% for male lone pensioners; the gender-gap becomes insignificant in northern Europe, with many countries demonstrating higher levels of fuel poverty amongst male lone-pensioner households. A remarkable 85.3% of male Portuguese lone pensioners and 88.9% of female lone pensioners are unable to afford adequate home heating, while Spanish and Greek lone pensioners face similar difficulties. Ireland appears to have the highest fuel poverty among lone pensioner households in northern Europe, with 11.8% for males and 7.8% for females. The UK fares badly in this respect. Corresponding statistics are 8.3% (male) and 7.3% (female) respectively, as does France (6% and 8.1% respectively). A key reason for many lone pensioners suffering fuel poverty is likely to be related to their financial situation, with many subsisting on very modest State pensions. Others may be living in older, less well-insulated dwellings and, thus, find it hard to heat the home even on less-modest pensions. It is likely that a combination of both factors is at play with this social group, although both factors are almost always necessary for the existence of fuel poverty.

House Type

It is useful to identify the type of accommodation where fuel poverty is highest. It might be expected that terraced and semi-detached houses would be, generally, better protected from the cold because the adjoining walls on either or both sides of the dwelling act as an insulator of heat. The results of this analysis do not necessarily corroborate such a hypothesis. In fact, the incidence of fuel poverty across Europe is highest in flat complexes (or multi-family dwellings), which indicates the existence of housing deprivation. Detached houses have a generally lower incidence of fuel poverty, especially in northern Europe where just 2.8% of such homes are declaring an inability to adequately heat the home.

Table 3.9 Households unable to heat home adequately by socio-demographic group (mean % of households, 1994-97)

	D	DK	NL	B	L	F	UK	IRL	I	EL	E	P	A	FIN
1 male aged –30	1.8	3.4	2.5	3.9	7.8	8.8	7.5	6.0	20.5	22.4	50.2	52.9	3.4	7.0
1 male aged 30-64	2.1	4.5	2.2	5.7	6.2	13.1	9.6	11.4	14.3	40.6	52.9	78.2	3.1	7.9
1 male aged 65+	1.4	3.6	2.9	3.6	2.0	6.0	8.3	11.8	25.6	62.6	67.7	85.3	0.7	3.6
1 female aged –30	2.6	4.1	1.7	3.3	4.9	9.8	6.6	9.7	20.6	29.7	50.9	56.3	3.9	10.5
1 female aged 30-64	3.9	6.8	4.6	7.5	3.4	12.3	10.2	9.8	20.3	52.6	54.4	79.4	2.8	7.2
1 female aged 65+	2.4	3.0	4.8	4.1	1.8	8.1	7.3	7.8	30.1	68.0	74.2	88.9	4.1	4.9
Single parent (All children <16)	2.4	3.9	12.0	6.9	9.3	15.0	18.6	19.3	22.9	48.5	62.7	71.4	3.3	8.8
Single parent (≥1 child 16+)	3.2	3.3	5.8	4.6	4.3	13.5	10.2	11.8	24.9	50.5	60.1	77.7	2.5	7.0
Couple without children (≥1 person 65+)	0.9	1.4	1.5	2.7	2.4	4.3	5.3	4.5	21.0	58.4	61.5	74.4	0.9	4.5
Couple without children (both <65)	1.6	1.9	0.9	3.1	2.2	6.0	4.4	3.9	14.8	41.1	45.6	64.7	1.4	3.8
Couple with 1 child	0.9	1.6	1.6	2.9	4.2	5.0	5.6	5.1	18.0	31.6	46.9	63.9	1.8	3.0
Couple with 2 children	1.1	1.3	0.9	3.9	1.0	4.8	3.7	3.8	18.4	33.9	46.0	63.1	1.9	3.6
Couple with 3+ children	1.6	2.2	0.3	3.6	6.0	6.4	9.5	8.3	32.0	41.2	58.5	84.2	4.4	2.8
Couple with 1+ child (≥1 child 16+)	2.3	1.7	1.1	3.2	1.7	7.6	5.1	4.0	22.2	42.7	51.2	68.8	1.4	3.9
Other households	2.1	2.7	5.0	5.0	0.4	6.9	9.2	7.0	23.3	47.4	58.2	79.0	1.3	2.9

Table 3.10 Fuel poverty in Europe by housing type (mean % of households, 1994-97)

	Detached	Semi-detached/ Terraced	Small MFD	Large MFD	Other
D	1.0	1.1	1.8	2.3	2.8
DK	2.8	4.1	3.6	2.9	4.5
NL	1.0	1.6	3.4	3.0	2.0
B	2.9	4.0	4.1	4.1	6.0
L	2.1	2.4	4.1	4.1	6.0
F	6.3	8.1	9.2	6.6	5.8
UK	2.4	6.3	11.2	9.2	5.6
IRL	4.7	7.4	14.1	17.4	18.0
I	26.3	11.1	25.4	13.9	35.9
EL	56.5	55.0	36.2	28.1	47.8
E	59.2	73.0	57.8	43.8	70.3
P	79.2	81.5	57.2	38.9	95.4
A	1.4	0.9	3.0	2.9	2.4
FIN	3.5	4.2	4.4	6.8	6.2

Note: MFD = Multi-family dwelling

Table 3.10 illustrates the results of fuel poverty across Europe by house type. Southern countries demonstrate the highest levels of fuel poverty across all house types. Portugal has the highest incidence of fuel poverty among detached, semi-detached and terraced households (with 79.2% and 81.5% affected). Spanish households living in small and large multi-family dwellings exhibit the highest incidence of fuel poverty, with 57.8% of households living in small apartment complexes and 43.8% of households living in large apartment complexes declaring fuel poverty. Detached, semi-detached and terraced houses are less susceptible to fuel poverty in northern Europe. However, the incidence of fuel poverty in large and small flat complexes in northern Europe is substantial. For apartment blocks with less than 10 dwellings, Ireland declares the highest levels of fuel poverty, with 14.1% affected, followed by the UK (11.2%) and France (9.2%). For large multi-family dwellings (apartment blocks with 10 or more units), Ireland, again, demonstrates the highest incidence of fuel poverty, with 17.4% affected – higher than the equivalent proportion of Italian dwellings of this type. The UK also has a relatively high incidence of fuel poverty for this dwelling type, with 9.2% declaring an inability to adequately heat their home. 'Other' households are also affected by fuel poverty, although this house type accounts for a small proportion of all households in the survey. It is likely that the high levels of fuel poverty in northern Europe among householders inhabiting apartments may be due to an income effect. Households in apartment blocks are more likely to be living on more modest incomes than those living in detached dwellings. As such, they may have limited disposable income available for home-heating purposes.

Marital Status

Marital status is a most interesting variable against which to analyse fuel poverty. As with poverty and deprivation research, it is likely that levels of fuel poverty may be highest among persons whose marital status is known to be associated with financial hardship. The separated and divorced, for example, are more likely to suffer from poverty than the married (Duncan *et al.*, 1993), and this is generally because such households are more likely to subsisting on a sole income. In addition, they may be lone-parent families raising children, compounding a potential (fuel-) poverty trap. The results show a marked pattern of fuel poverty across Europe, with the highest levels found among separated households. An EU-14 mean incidence of fuel poverty of some 33.9% is calculated for this household type, followed by 21% of widowed households. If the analysis is shifted to EU-10 (non-Mediterranean countries), an average incidence of fuel poverty of 13.1% of separated households is reported. Widowed persons suffer to a far less extent in northern Europe, however divorced households appear to suffer from fuel poverty, with 7.1% such households in northern Europe declaring an inability to afford adequate household warmth. Conversely, married households suffer proportionately less than other households, with incidences in EU-14 and EU-10 of 15.6% and 2.9% respectively.

Separated, Divorced and Widowed It is clear that fuel poverty is strongly related to poverty, and, as such, it is no surprise that low-income households declare the highest levels of fuel poverty across Europe. While sample sizes for separated, divorced and widowed households were too small to calculate reliable estimates for certain countries, it is clear that separated households in southern Europe demonstrate alarming levels of fuel poverty, with a remarkable 81.9% affected in Portugal. Spanish separated households do not fare well either, with an incidence of 66.9%, while the proportions of separated households in Greece (48.5%) and Italy (21.4%) are higher than any levels found in northern Europe. There are, however, some notably high incidences among separated households in EU-10, most especially in Ireland where 18.3% are suffering. Second, divorced households in southern Europe demonstrate very high levels of fuel poverty, with Portugal most affected (81.9%), followed by Spain (66.9%) and Greece (48.5%). Italy's level of fuel poverty among divorced households (12.7%) is less than the UK's level of 13.7% which appears to be the highest in northern Europe. Third, widowed households are suffering from fuel poverty, though not, it appears, in northern Europe where low incidences are calculated. Some 83.9% of widowed households in Portugal, 63.6% of those in Spain and 55.9% of those in Greece are declaring fuel poverty. Taken as a whole, these results indicate that marital status strongly affects the likelihood of fuel poverty. Married households are found to have mostly negligible levels of fuel poverty outside southern Europe.

Table 3.11 Fuel poverty in Europe by marital status (mean % of households, 1994-97)

	Married	Separated	Divorced	Widowed	Never married
D	1.2	-	3.5	2.1	1.6
DK	2.1	-	5.1	3.5	3.1
NL	1.0	-	5.7	4.4	1.3
B	2.9	8.8	6.1	4.0	3.8
L	2.5	-	-	-	4.2
F	5.0	14.0	12.4	7.5	8.7
UK	4.5	11.2	13.7	8.2	6.3
IRL	4.7	18.3	-	7.0	6.2
I	20.7	21.4	12.7	25.8	24.6
EL	45.5	48.5	49.1	55.9	43.3
E	52.4	66.9	58.0	63.6	53.9
P	71.6	81.9	67.7	83.9	74.7
A	1.4	-	3.2	2.5	2.1
FIN	3.6	-	6.4	4.8	4.9

Highest Educational Attainment

Educational attainment is, in general, a good indicator of household income and social class and poverty research has consistently shown a strong association between low education levels and high levels of poverty and deprivation. This study tests this hypothesis with fuel poverty. The results demonstrate a linear relationship between educational attainment and fuel poverty both in northern Europe and across EU-14. Average incidences of fuel poverty of 7.4% are found across Europe for those with third-level qualifications. This increases to 12% among those who completed their secondary schooling, and 19.2% among those households who did not complete their second-level education. In northern Europe, a similar pattern is found, although the incidences are far lower. A negligible level of fuel poverty of just 2.2% is reported amongst households with third-level qualifications, followed by 3% among those who completed secondary school, while an incidence of fuel poverty of 4.7% of households is found for households that finished their formal education before completion of secondary school.

Table 3.12 Fuel poverty by educational attainment (mean % of households, 1994-97)

	Third Level	Upper Secondary	Lower Secondary
D	1.9	2.1	3.4
DK	2.0	2.4	3.8
NL	0.7	1.2	2.7
B	2.6	3.4	4.0
L	3.7	2.3	2.8
F	3.9	5.9	8.2
UK	2.4	4.6	8.3
IRL	1.8	3.2	8.4
I	7.9	14.7	25.5
EL	21.4	32.7	54.9
E	29.6	40.9	62.5
P	21.9	49.7	79.0
A	1.7	1.4	2.2
FIN	2.8	4.5	5.0

Low Educational Attainment Comparatively, the results show some interesting patterns across Europe. In northern Europe incidences of fuel poverty among households with low levels of educational attainment are relatively high in Ireland, the UK and France, with levels of 8.4%, 8.3% and 8.2% respectively. Levels in southern Europe are dramatic, with some 79% affected in Portugal, 62.5% in Spain and 54.9% in Greece (Table 3.12). Conversely, households with a third-level qualification in northern Europe appear to suffer negligible rates of fuel poverty. However, moderate to high incidences are found in some southern countries. Spain reports an incidence of some 29.6% among households with high levels of educational attainment, while Portugal and Greece perform similarly poorly in this regard (21.9% and 21.4% respectively). The results again point to a strong relationship between fuel poverty and household income.

Main Income Source

It is very useful to examine fuel poverty by households' income sources, as they are a good indication of income level. Dependence on unemployment assistance, state pension or other social transfers implies that such households live on a modest level of household income. Such income may preclude households from heating their home adequately in a direct sense through not being able to afford fuel bills, or indirectly through being unable to invest in household energy-efficiency improvements. The unemployed and those on other forms of social-welfare payments demonstrate the highest incidences of fuel poverty in Europe. Some 27.7% of those on social welfare and 26.3% of those unemployed are demonstrating fuel poverty overall, while in northern Europe the respective proportions are 10.2% and 12.1%. Self-employed and waged households generally display low or negligible levels of fuel poverty across northern Europe.

Table 3.13 Fuel poverty in Europe by main income source (mean % of households, 1994-97)

	Wages (employee)	Self-employed /Farming	Pensions	Unemployed	Other social transfers	Private
D	1.3	1.0	1.8	4.9	6.6	0.8
DK	2.1	2.5	3.4	10.3	4.6	1.5
NL	0.7	0.6	3.1	5.0	6.5	1.4
B	2.8	3.7	3.2	9.5	7.4	5.4
L	3.3	1.8	1.9	-	8.2	-
F	6.0	6.1	6.1	21.3	-	8.9
UK	3.2	3.0	5.9	17.7	19.5	3.5
IRL	3.4	2.9	7.5	23.3	16.3	6.8
I	18.4	17.2	24.9	46.6	17.4	28.8
EL	36.5	41.5	58.3	-	44.0	37.5
E	47.6	44.6	65.3	79.2	65.5	52.6
P	71.6	64.6	81.8	81.1	73.9	56.0
A	1.7	1.1	2.2	9.9	84.5	6.5
FIN	4.1	2.9	5.0	4.9	3.7	7.6

The Unemployed and Social-Welfare Recipients The overall highest incidence of fuel poverty in the unemployed category is demonstrated in Portugal, where 81.1% are considered fuel-poor, as is shown in Table 3.13. Elsewhere in southern Europe, Spain (79.2%) and Italy (46.6%) also show high levels. In northern Europe 23.3% of Irish unemployed households are suffering from fuel poverty. Likewise, there are high incidences in France amongst the unemployed (21.3%) and the UK (17.7%) (Table 3.13). For those whose main income source is other forms of social welfare, high levels are found, again, in southern Europe. Portugal (84.5%), Spain (73.9%), Greece (65.5%) and Italy (44.0%) all demonstrate high incidences of fuel poverty. In northern Europe, the highest incidence of fuel poverty among this group of social-welfare recipients is found in France (19.6%), followed by Ireland (17.4%) and the UK (16.3%).

Housing Tenure

Housing tenure is an important dynamic of fuel poverty, as has been noted by Whyley and Callender (1997). This is because it gives households varying levels of control over their home, heating systems and their energy consumption. Owner-occupiers may be considered as fully autonomous, while tenants may be more limited as regards what they feel they can afford to do to improve their housing or even what they are authorised to do to improve their housing (Clinch and Healy, 2000a).

Table 3.14 Fuel poverty in Europe by housing tenure (mean % of households, 1994-97)

	Owner	Tenant	Rent-Free
D	0.8	2.3	0.7
DK	2.2	4.8	4.5
NL	0.5	3.7	1.9
B	2.9	5.4	2.7
L	2.3	5.0	-
F	5.6	9.6	7.1
UK	3.1	11.9	5.6
IRL	4.1	20.9	9.2
I	19.9	27.3	22.8
EL	49.3	52.8	51.3
E	52.0	67.3	67.2
P	71.8	77.8	83.4
A	1.1	3.5	3.0
FIN	3.8	8.6	6.0

Tenants Table 3.14 illustrates the results for this section. Owner-occupiers suffer least from fuel poverty, while tenants suffer most. Those whose accommodation is provided free of charge by the State suffer above average. The mean incidence of fuel poverty for tenant households across all of Europe is found to be 21.5%, while in northern Europe the corresponding proportion is 7.6%. In southern Europe rates of fuel poverty among tenant households are very high, with levels of 77.8% in Portugal, 67.3% in Spain, 52.8% in Greece and 27.3% in Italy are all above the average (EU-14) level. In northern Europe, the highest incidence among tenants are found in Ireland, where a remarkable 20.9% of tenant households declare fuel poverty. The UK (11.9%) and France (9.6%) also suffer relatively highly in this regard (Table 3.14).

Comparison with Traditional Definition

This section compares results using the ECHP data with those found using the standard (Boardman) definition of fuel poverty in which a household is considered fuel-poor if it would require more than 10% of its household income to heat its home. The results for the UK are set out in Table 3.15. It should be borne in mind that the ECHP data demonstrate lower bound levels of fuel poverty for northern Europe and upper bound estimates for southern Europe because of the phrasing of the key question relating to fuel poverty, ability to heat the home adequately. Households in southern Europe that suffer occasional or intermittent problems in affording adequate home heating are being categorised in the same group (fuel-poor) as those in colder northern European countries where more chronic fuel poverty may exist.

Table 3.15 UK fuel poverty in 1996: a comparison using alternative approaches

	ECHP (1996)	EHCS (1996)
Overall	5.3	22.2
Owner-Occupier	2.6	16.7
Rent (privately)	13.4	39.0
Rent (public / local authority)	12.3	35.0
Lone Pensioner	7.2	36.3
Lone Parent	13.5	10.5
Couple 65+, no children	3.4	13.6
Couple with children	2.7	6.1

Nonetheless, the comparison uses 1996 data from the ECHP and the equivalent data from the 1996 English Housing Conditions Survey (EHCS) reported in DEFRA and DTI (2001). The table clearly indicates that levels of fuel poverty are considerably lower using the ECHP data than using the traditional EHCS data. The national estimate is just 5.3% using the ECHP indicator compared with 22.2% using the EHCS data. The only instance in which data from the ECHP indicate higher incidences than the EHCS findings is in regard to the level of fuel poverty among lone parents. Some 13.5% of lone parents are declaring fuel poverty in the ECHP survey compared with 10.5% using the EHCS data. The comparison highlights the mismatch found between the two approaches. The ECHP almost persistently indicates that the incidence of fuel poverty is lower than that which has been traditionally reported using the standard Boardman definition based on fuel expenditure.

The lower levels found using the experiential ECHP data provide grounds for arguing that the Boardman definition of fuel poverty should be, at the very least, altered to reflect the ostensibly smaller magnitude of the problem, particularly at the national level. However, care should be taken in interpreting the data. It is important to re-state that, because of the phrasing of the questions relating to fuel poverty (and because of other exogenous factors such as climate), the ECHP data report lower bound estimates in northern Europe and upper bound estimates for southern Europe, and this accounts for at least some of the apparent mismatch found in the levels of fuel poverty reported in Table 3.15.

Multivariate Regression Analysis

It is useful to conduct more rigorous statistical and econometric analysis of the results to test the strength of the significance of the results. In this regard, this section presents a multivariate Probit regression analysis which examines those factors that influence the probability of being fuel poor. A number of indicators of fuel poverty are regressed against socio-economic and other characteristics of households. A Probit analysis allows us to examine the household characteristics that are significantly associated with each indicator of fuel poverty. The added value over the bivariate analysis presented in the cross tabulations is that we can examine the effect of each variable holding all else equal. The results for the earlier waves of the survey are generally more robust. This is most likely because of the larger sample size. Overall, the predictive power of the models is good and it is reassuring that, in general, the same variables remain significant across the four waves.

The results of the Probit model for the household's ability to heat the home adequately, the main indicator of fuel poverty, are set out in Tables 1 and 2 of Appendix I in this book. The results suggest that, *inter alia*, being younger, the number of children, marital status, income source, housing tenure, being in receipt of a housing allowance, having poor health status and being less well-educated are significantly associated with being unable to heat the home adequately. The marginal effects suggest that age, marital status, health, education and being in receipt of a housing allowance are most strongly related to the inability to heat the home adequately.

The results of a Probit model for the household's inability to pay utility bills in the last twelve months are set out in Tables 3 and 4 of Appendix I of this book. The results suggest that, *inter alia*, being younger the number of children, marital status, being unemployed or on benefit, living in rental or local authority accommodation, having poor health status and being less well educated are significantly associated with being unable to pay bills in the last twelve months. Interestingly, those households in receipt of a housing allowance are significantly less likely to be unable to pay their bills. The marginal effects suggest that age, income source and education are most strongly related to being unable to pay bills in the last twelve months.

Results showing those factors that influence the probability of a household having inadequate heating facilities are set out in Tables 5 and 6 of Appendix I in this book. The results suggest that, *inter alia*, being younger, household composition, marital status, tenure, accommodation type, having poor health status and being less well-educated are significantly associated with the respondents' perception of the lack of adequate heating facilities. The marginal effects suggest that health status and education are most strongly related with the respondent's perception of the lack of adequate heating facilities.

The results of a Probit model examining the probability of the presence of central heating in the household are set out in Tables 7 and 8 of Appendix I. The results suggest that, *inter alia*, being younger, household composition, marital status, being unemployed or on benefit, living in rental or local authority accommodation, having poor health status, being less well-educated and being in receipt of a housing allowance are significantly associated with a lack of central heating in the accommodation. The marginal effects suggest that marital status, income source, housing tenure, health status and education are most strongly related to the absence of central heating in the accommodation.

Results showing those factors that influence the probability of the presence of damp in a household are presented in Tables 9 and 10 of Appendix I. The results suggest that, *inter alia*, being younger, the number of children, marital status, being unemployed or on benefit, living in rental or local authority accommodation, accommodation type, having poor health status and being less well-educated are significantly associated with the presence of damp in the respondents' accommodation. The marginal effects suggest that health status, tenure, accommodation type and education are most strongly related to the presence of damp.

Finally, the results of a Probit model for the presence of rot are presented in Tables 11 and 12 of Appendix I. The results suggest that, *inter alia*, the number of children, marital status, being unemployed or on benefit, living in rental or local authority accommodation, having poor health status and being less well educated are significantly associated with the presence of rot in the respondent's accommodation. The marginal effects suggest that health status, tenure and education are most strongly related to the presence of rot.

Cooling Requirements

Some words on *affordable cooling* are now warranted. It is estimated that between 15,000-20,000 people, most of them elderly, may have died during a heatwave that blanketed Europe during the summer of 2003. France was hit hardest by the widespread health emergency sparked by two weeks of stifling temperatures, with the country's largest undertakers' group putting the death toll at about 10,000. In Spain, officials have said that more than 100 people died in the first two weeks of August, when thermometers soared daily to 40 degrees Celsius across large swathes of Europe; the latest estimate has claimed that 2,000 people may have succumbed to the intense heat. According to the latest provisional data, Portugal reported 1,316 more deaths across the country between July 30 and August 12 2003, as compared with the same period last year, and this may be attributable to the soaring mortality rate from the heat. In the UK, where the hottest temperatures on record were obtained in summer 2003, the Office for National Statistics said that 907 additional deaths were registered in the week ending August 15 as compared with 2002. Press reports in Italy have put the death toll at 1,000, although no official estimates have been issued. In the Netherlands, officials say between 500 and 1,000 more people died in July and August than during an average summer, but figures were deemed preliminary. Germany was largely spared by the heatwave, with only about three-dozen deaths blamed on the unusually hot weather. Precise figures have been difficult to obtain, as many countries do no collect specific data on heat-related deaths.

It is certainly reasonable to hypothesise that some low-income households in southern Member States are less likely to afford air-conditioning units and other forms of expensive cooling systems and this in turn results in summertime energy poverty. Judging by the very tentative and provisional vital statistics data that have been released by governments across Europe following the summer heatwave, it is probably equally reasonable to state that there are some excess summer deaths and perhaps other adverse health outcomes as a result of such energy poverty. However, this thesis – though undeniably fascinating – is outside this study's terms of reference. Poor thermal efficiency is not going to be a factor in any health-related impacts of excessive heat exposure. In addition, and more importantly, there are no questions in the ECHP datasets regarding adequate household cooling, though it is suggested that future ECHP surveys incorporate a variable to capture affordable summertime home cooling as well as affordable wintertime home heating.

The link between summertime energy poverty (itself not a widely known term) and excess summer mortality (ditto) has not been formally addressed hitherto and it will take some years before enough data are available to test the relationship and provide epidemiological evidence. The increasing likelihood of future summer heatwaves (through continued global warming and climate change) will provide grounds for this topic to be given more attention through increased research funding. However, it is important to stress that, in the absence of longitudinal data on summertime energy costs and excess summer deaths, the matter is likely to remain unresearched; and such hypotheses, though likely to be proved true in time, cannot be tested rigorously until a time-series has arisen.

Conclusions

This chapter contains the first cross-country analysis of fuel poverty using comparable data. A new methodology for measuring fuel poverty has been proposed which attempts to identify the fuel-poor using a number of social indicators of deprivation and a composite measurement. The data employed are longitudinal and taken from the first four waves of the European Community Household Panel (ECHP). It is argued that the proposed new methodology for measuring fuel poverty is superior to the problematic and unscientific Expenditure approach for measuring fuel poverty (households spending more than 10% of income on home heating), especially in cross-country comparisons where such a measurement may be rendered meaningless because of differing purchasing power and varying real energy prices. These results compare favourably with those reported by Whyley and Callender (1997) in their provisional analysis of fuel poverty using social indicators.[8] Using six objective and subjective indicators in a weighted composite index, fuel poverty is calculated to be highest in southern Europe. Portugal, Greece, Spain and Italy demonstrate the highest levels of fuel poverty, regardless of sensitivity analysis. In northern Europe rates of fuel poverty

---

[8] A composite index was not constructed by Whyley and Callender.

are lower, however France, Belgium, the UK and Ireland exhibit relatively high incidences.

Using an Expenditure approach, it has been calculated that 16.4% of British households are fuel-poor (DEFRA and DTI, 2001). Similar results are found for Ireland: between 20.7% (housing costs included) and 25.0% of homes are found to be suffering from fuel poverty using the standard definition. These estimates are higher than the results based on social indicators in this research, and it is argued that the Expenditure approach may over-estimate actual levels of fuel poverty because the arbitrary 10% income threshold may be too low. As the quantitative literature on fuel poverty is undesirably thin, there has been little theoretical debate about the appropriate income threshold. It is suggested that this matter be given attention in future research so that both approaches may be consolidated.

Single-parent households are shown to have the highest incidence of fuel poverty across Europe, especially in southern countries where as much as three-quarters are fuel-poor. In northern Europe the highest incidence is found in Ireland and the UK where about one-fifth of lone-parent households are suffering fuel poverty; such results are very high relative to most northern countries. Once again, there is a considerable variation in the results when compared with the standard approach to calculating fuel poverty (just 11% of lone-parent households in the UK are fuel-poor using the traditional definition). Lone pensioner households are also identifiable as a risk group, especially in southern Europe where up to 88.9% of such households are declaring fuel poverty. In northern countries Ireland and the UK, again, display the highest levels in this social group. Households living in apartment complexes also exhibit high incidences of fuel poverty, especially in Ireland, the UK and southern countries. The unemployed are a key risk group, with 81.1% affected in Portugal. In northern Europe the unemployed in Ireland, France and the UK suffer highest, with 23.3%, 21.3% and 17.7% affected respectively. Tenants are at a far greater risk of fuel poverty than owner-occupiers across Europe, as are those with low educational attainment. Finally, the divorced, separated and widowed display persistently high incidences of fuel poverty in Europe. All of these results are highly significant in the multivariate Probit model.

In the context of recent and ongoing fuel crises, fuel poverty is an issue which, despite some government intervention, has not disappeared. However, the longitudinal results in this chapter show some improvements for certain countries affected by fuel poverty, including the UK and Ireland. Perhaps the key result of interest in this analysis is the astonishingly high rates of fuel poverty found in southern Europe, especially in Portugal, Spain and Greece. It is clear from the socio-economic analysis that, throughout Europe, fuel poverty is confined to very specific, discrete social groups. These groups should be targeted for subvention to upgrade their housing and retrofit energy-saving technologies so that such hardship may be alleviated, or at least reduced. In northern Europe, where incomes are substantially higher, it is clear that there remains a significant information gap regarding the benefits of energy-efficiency measures, and high-income households need to be persuaded to retrofit using their own funds, while the unemployed and households comprised of pensioners and lone parents should be targeted for large

State-subsidised energy-efficiency retrofit programmes. Similar recommendations were made in a previous analysis of fuel poverty in Ireland (Clinch and Healy, 2000a). It is clear from some of the longitudinal results presented in this chapter, including those relating to the UK and Ireland, that such a policy mix can reduce fuel poverty substantially.

Chapter 4

The Severity of Fuel Poverty in Ireland

Introduction

The previous chapter analysed cross-country levels of fuel poverty. However, it is thought beneficial to analyse the case of Ireland in more detail for a number of reasons, particularly because of its outlier status in the results. A large household survey has been developed and employed to assess the severity of fuel poverty in Ireland, an interesting country to examine for four key reasons. First, very little empirical research on fuel poverty exists in Ireland because of the lack of suitable data hitherto. Second, Ireland has been identified as a country marked by among the highest levels of housing deprivation, and among the least energy efficient dwellings in northern Europe (Chapter 4). Third, Ireland, like the UK, has among the highest levels of seasonal variations in mortality, leading many researchers to believe that the relatively poor thermal efficiency of the Irish housing stock is a major reason for these rates of mortality (Eng and Mercer, 1998). Fourth, Ireland's spectacular economic success over the last decade has simultaneously improved the quality of the housing stock and placed a considerable burden on policy-makers to achieve various challenging environmental targets, most notably on emissions of greenhouse gases and acidification precursors. By delivering the first estimates of fuel poverty in Ireland and its evolution over time, this study assists policy-makers in assessing how much the alleviation of fuel poverty would make in bridging the gap between business-as-usual energy-related environmental emissions and emissions from an energy efficient domestic sector.

    The study contributes to the methodological discussion on fuel poverty by distinguishing long-term (chronic, persistent) sufferers from short-term (intermittent, occasional) sufferers. The recent (2001) survey data also allow for a great degree of data disaggregation, and a detailed socio-economic and socio-demographic analysis pin-points those suffering disproportionately from fuel poverty in Ireland. Moreover, the relationship between fuel poverty and adverse housing conditions, such as damp and condensation, is examined, while the effect of fuel subsidies is also analysed. The survey also identifies the key reasons for non-investment in energy-saving measures by Irish households. This is a crucial component of the study as it identifies the relative importance of the various reasons for market failure in domestic energy efficiency, outlined in Clinch and Healy (2000a), and thus contributes strongly to the policy debate on energy efficiency and fuel poverty. Some discussion is given to the various environmental, public-health and social-policy implications of these findings. The chapter begins with the results of the survey regarding the thermal efficiency of the dwelling stock. Data from two other previous surveys of the housing stock are compared to assess the penetration of energy-efficiency measures over time.

Domestic Energy-Efficiency Levels

The data in this study are derived from a statistically representative, face-to-face national household survey of Ireland, conducted in spring 2001. The sample was selected randomly using probability-based sampling. Results are only reported in this study where sample sizes allow for a reasonably low level of error. Results based on sample sizes of less than 30 households are not reported in Tables, Figures or in the text itself. An effective sample of 1,500 households was achieved using a probability-based sampling procedure and the survey was carried out by a professional survey unit. To test for non-response bias, four key variables from the sample (age, sex, marital status and economic activity) were compared with corresponding Irish census estimates. The characteristics of the sample are broadly similar to those of the Irish adult population. Some variation is to be expected given the last census was carried out in 1996. Given the broad 'representativeness' of the sample, no corrective weighting procedures were applied to the data.

The overall level of fuel poverty is directly related to the level of household income and energy efficiency (Boardman, 1991). Previous research by this author has indicated that Ireland and the UK have similarly poor levels of domestic energy efficiency (Healy, 2001a). The first results of the survey report the latest levels of ownership of various energy-saving measures in the dwelling stock. To assess the penetration of various measures over time, two previous surveys of the dwelling stock's energy-efficiency standards are compared.[1] The results are shown in Table 4.1.

Table 4.1 Percentage of Irish households with various energy-saving measures (1996-2001)

	1996	1998	2001
Lagging jacket	-	64	86
Floor insulation	22	24	25
Roof insulation	72	72	78
Wall insulation	42	42	42
Double glazing	33	37	64
Draught stripping	-	37	40
Low-energy light bulbs	-	-	29
Central heating	74	80	86

Source: 1996 data from Eurostat (1999); 1998 data from Clinch and Healy (1999a).

---

[1] The 1996 survey relates to a pan-European study conducted by the European Statistical Office (see Eurostat, 1999), while the 1998 data are taken from an Irish energy-efficiency study (see Clinch and Healy, 1999).

The penetration of hot-water cylinder lagging jackets has increased dramatically (by over a third) from 1998 to 2001, with 86% of the dwelling stock (1.3 million dwellings) now equipped with this measure. Levels of floor insulation have remained relatively static over the past 5 years, with just a quarter of Irish houses so equipped. The penetration of roof insulation is good in Ireland, with almost four-fifths of the stock possessing this energy-efficiency measure. Much of this success is due to the State-funded attic-insulation scheme of the 1980s, but the improvement since 1996 is likely to be a product of rising real incomes. The most significant improvement since the 1998 study relates to improvements in double glazing. Over a quarter if the housing stock has retrofitted double glazing over the past four years; again, it is likely that rapidly rising living standards were a causative factor behind this success. Levels of cavity-wall insulation in Ireland are low (42%) and remain static over the period 1996-2001.[2] Draught stripping of doors and windows also remains similarly low (40%). This survey produces an estimate for low-energy light bulbs, and reports that about one-in-three households own such measures. Finally, the penetration of central heating has increased from 74% in 1996 to 86% in 2001, representing a 16% increase over five years. As can be seen, there is considerable scope for improving the energy-efficiency standards of the Irish dwelling stock, especially with regard to floor and wall insulation and draught stripping of doors and windows.

Aggregate (National) Fuel Poverty by Severity

This section presents the household survey's aggregate results for fuel poverty by severity of experience. It is very useful to identify chronic, persistent, long-term fuel-poverty sufferers against those whose problems are more short-term, occasional or intermittent.[3] To address this important issue, the survey asked all households in the sample to state how often they were unable to adequately heat their home – the basic qualitative definition of fuel poverty first stated by Lewis (1982) – on a four-point response variable. The results, reported in Table 4.2, show that some 12.7% of Irish households suffer occasional fuel poverty (amounting to some 165,000 households), while a further 4.7% suffer persistent fuel poverty (or 62,000 households). Therefore, a total of 17.4% of Irish households are declaring some level of fuel poverty in 2001, which amounts to approximately 226,000 households.

This result is cross-checked against the standard quantitative definition of fuel poverty where households are classified as fuel-poor if they would need to spend more than 10% of their income on energy in the home (Boardman, 1991). Between 20.7% (housing costs included) and 25.0% of homes are found to be

---

[2] Not all housing can be fitted with cavity-wall insulation, as some housing is of the solid-block type (e.g. Georgian, Victorian period).
[3] Note that the terms 'chronic', 'persistent' and 'long-term' are used interchangeably throughout the thesis, as are 'short-term', 'intermittent' and 'occasional'.

suffering from fuel poverty using this definition.[4] Such results are similar to those found recently in the UK of 16.4% (housing costs included) to 22.3%.

Table 4.2 Ability to adequately heat home: incidence and numbers affected (Ireland, 2001)

	% of households	# households
Some difficulties ('intermittent')	12.7	165,000
Usually not	2.5	33,000
Never	2.2	29,000
'Chronic' fuel-poor (2+3)	4.7	62,000
Total fuel-poor (1+2+3)	*17.4*	*226,000*
'Standard' definition (Housing costs included)	20.7	269,000
'Standard' definition (Housing costs deducted)	25.0	325,000

This aggregate level of fuel poverty in Ireland is higher than the incidence produced in Chapter 3 using the European Community Household Panel (ECHP) which contains a binary-response (Yes/No) question regarding the ability to adequately heat the home. Estimates of 8.0%, 5.9%, 6.5% and 5.1% were produced for the years 1994-97 inclusive. If the continuing downward trend persisted over the years 1998-2000 (in line with rising living standards and real increases in net household income), then it is suggested that the chronic estimate of fuel poverty of 4.7% reported here corresponds well with the results found in Chapter 3. It is believed that a large portion of the intermittently fuel-poor are not declaring problems of fuel poverty in the ECHP data from the 1990s which appears to capture *persistent*, as opposed to *intermittent*, fuel poverty and, as such, a lower incidence is reported using the binary-response variable.

Fuel Poverty by Socio-Demographic Group

Some households need more fuel than others because their circumstances dictate that the home must be heated for longer intervals or because they require higher temperatures (e.g. households occupied by the elderly or those with very young children). For some households, heating costs may be disproportionately high because these costs may fall on one person (e.g. single-adult households). Furthermore, there are some households that are known – through generalised poverty and deprivation research – to face severe financial hardship (e.g. single-parent households). All of these factors affect the probability of certain household groups enduring fuel poverty. The socio-economic and socio-demographic analysis in this section examines a variety of household types based on demographic (e.g.

---

[4] This is based on 1994/95 Household Budget Survey data.

marital status, educational attainment) and economic (e.g. household income, social class) variables.

Marital Status

Marital status is a most interesting variable against which to analyse fuel poverty. As with poverty and deprivation research, it is likely that levels of fuel poverty may be highest among persons whose marital status is known to be associated with financial hardship. The separated and divorced, for example, are more likely to suffer from poverty than the married (Duncan et al., 1993), and this is generally because such households are more likely to be subsisting on a sole income. In addition, they may be lone-parent families raising children, compounding a potential (fuel-) poverty trap.

Table 4.3 Ability to heat the home adequately by marital status (% households)

	N	Some difficulties	Usually not	Never	Total % fuel-poor
Single	522	15.3	4.8	2.5	22.6
Married / Co-habiting	809	9.0	1.2	1.7	11.8
Separated / Divorced / Widowed	149	22.8	1.3	4.0	28.2

The highest incidence is found amongst separated, divorced and widowed households, with 28.2% affected by fuel poverty, followed by single persons, with an incidence of 22.6% (Table 4.3), similar to the results of DEFRA and DTI (2001) for the UK. Single people also account for the highest group of fuel-poverty sufferers in absolute numbers in Ireland, with 104,000 households affected, compared with 85,000 married or co-habiting households and 37,000 widowed, divorced or separated households. In terms of severity, single people appear to have the highest levels of chronic fuel poverty, with 32.3% of fuel-poverty sufferers declaring persistent fuel poverty, compared with 19.1% of those householders widowed, divorced and separated.

Lone–Versus–2-Parent Families

In order to identify the levels of fuel poverty among lone parents, households in the survey sample are split so that only those with dependent children are included. Two sub-samples are analysed: those single, separated, divorced or widowed (collectively denoted as lone-parent families) and those married or co-habiting. The results are alarming and show that single-parent households are demonstrating levels of fuel poverty of almost four times that found amongst two-parent families. An overall incidence of 40.2% is quantified for single-parent families, which represents approximately 29,000 households nationally. About 15.2% of lone-parent families in fuel poverty are demonstrating chronic levels (Table 4.4). This incidence is far higher than that reported for the UK by DEFRA and DTI (2001), which showed a rate of 15% for lone parents with dependent children, indicating that Irish single-parent families are suffering disproportionately highly from fuel poverty.

Table 4.4 Ability to heat the home adequately by marital status of parents (% households)

	N	Some difficulties	Usually not	Never	Total % fuel-poor
Single / Separated / Divorced / Widowed	82	34.1	4.9	1.2	40.2
Married / Co-habiting	479	7.7	1.3	1.9	10.9

Spatial Distribution

There is very little difference in the incidence of fuel poverty by region: between 15% and 18.9% of households are identifiable as either occasional or chronic fuel-poverty sufferers (Table 4.5). However, a significant result is found in absolute numbers, where a U-shaped result appears with high numbers of fuel poverty found in very urbanised and very rural areas. Some 87,000 rural households and over 63,000 households in Dublin City report an inability to heat the home compared with 22,000 in large towns and 26,000 in cities outside Dublin. The highest incidence of chronic fuel poverty is found in small towns (7.8%).

Table 4.5 Ability to heat the home adequately by region (%)

	Dublin City	Other City	Large Town (10-40,000)	Small Town (<10,000)	Village/ Rural
N	371	162	160	184	565
Some difficulties	13.7	14.8	11.3	10.4	12.7
Usually not	2.4	1.9	3.1	2.3	2.5
Never	2.7	1.2	0.6	4.5	1.8
Total % fuel-poor	18.9	17.9	15.0	17.1	17.0

Educational Attainment

There is a very strong, negative, linear relationship between educational attainment and the incidence of fuel poverty. The highest level of fuel poverty is found in those groups with low levels of educational attainment (Table 4.6). For instance, primary school-leavers have an incidence of some 25.6%. In addition, those who left secondary school before Leaving Cert/Tech-equivalent (i.e. final secondary school examinations), have high levels of fuel poverty (incidence of 21.9% and 55,000 households, making this group the largest in absolute terms). Conversely, those with third-level qualifications have very low levels of fuel poverty; only 3.6% of those households with a primary or postgraduate degree suffer fuel poverty, which amounts to fewer than 2,000 households nationally. Persistent fuel poverty is highest proportionately among those who left school at Tech/VOC-level, with 38.7% of fuel-poor households in this group suffering chronic fuel poverty.

Table 4.6 Ability to heat the home adequately by educational attainment (% households)

	N	Some difficulties	Usually not	Never	Total % fuel-poor
No formal education	4	-	-	-	-
Primary	203	18.7	4.4	2.5	25.6
Lwr. secondary	278	16.5	2.9	2.5	21.9
Upr. secondary	187	18.7	1.1	1.6	21.4
Upr. sec. Tech / VOC	424	8.7	2.4	3.1	14.2
Upr. sec. Leaving Cert.	88	10.2	3.4	2.3	15.9
Upr. sec. both Tech & Leaving Cert.	108	10.2	0.9	1.9	13.0
3rd level non-degree	94	7.4	1.1	0	8.5
3rd level primary, professional or postgraduate	56	1.8	1.8	0	3.6

Dwelling Age

Data relating to dwelling age has been grouped into four discrete categories on the basis of energy-efficiency characteristics. Pre-1940 dwellings (which account for about 13% of all dwellings in the sample) were mainly solid-wall construction, while during the 1940s through to the 1970s, cavity-wall construction was implemented. During the 1980s, improved U-values for both walls and attics were introduced in various building regulations, increasing the thermal efficiency of the dwelling, and these U-values have been further enhanced in the 1990s with the introduction of more stringent building regulations, especially those in 1997. Improvements in housing construction are reflected broadly in the results, which show that older dwellings are more likely to be fuel-poor than newer dwellings, with the highest incidence of fuel poverty among homes built before 1940 (18.5%) as can be seen in Table 4.7.

These results are similar to those found in the UK (DEFRA and DTI, 2001). As might be expected, those in newer homes suffer less from fuel poverty, both proportionately and in absolute numbers, although the level of fuel poverty among homes built since 1990 (at 13.6%) is, perhaps, higher than might be expected. It is likely that a large portion of these households are struggling to keep their homes adequately heated for financial reasons, as many of these houses are likely to be first-time home-owners and are finding it difficult making ends meet. Houses built between 1941 and 1979 demonstrate the highest level of persistent fuel poverty; some 31.9% of fuel-poor households in this group declare chronic fuel poverty.

Table 4.7 Ability to heat the home adequately by dwelling age (% households)

	N	Some difficulties	Usually not	Never	Total % fuel-poor
Pre-1940	173	13.9	2.3	2.3	18.5
1941-79	697	11.3	2.4	2.9	16.6
1980-89	284	9.9	1.1	1.4	12.4
1990 to date	206	11.1	1.0	1.5	13.6

Occupancy

There is a very strong, negative relationship between household occupancy levels and fuel poverty. The highest levels of fuel poverty (both in incidence and in absolute numbers) occur in smaller households, especially one- and two-person occupied houses (Table 4.8).

Table 4.8 Ability to heat the home adequately by housing occupancy (number of persons) (% households)

	N	Some difficulties	Usually not	Never	Total % fuel-poor
1	204	19.6	4.4	4.4	28.4
2	297	13.8	4.0	3.0	20.9
3	222	14.9	0.5	2.3	17.6
4	378	9.3	2.4	1.1	12.7
5	262	9.5	1.5	0.4	11.5
6+	115	9.6	1.8	4.3	15.7

Some 106,000 one- and two-person households are declaring either intermittent or persistent fuel poverty, and almost three-in-ten single-person occupied households are suffering. However, the results also point to a U-shape relationship between household size and fuel poverty, with an increasing incidence (15.7%) in very large households (those with six or more persons). Chronic fuel poverty is proportionately highest in households with six or more occupants. Some 38.9% of fuel-poor households in this group are persistently fuel-poor. These results are similar to those found in the UK (DEFRA and DTI, 2001) where one-person households and large adult households were found to be most at risk.

Fuel Poverty by Socio-Economic Group

This section analyses fuel-poor households by various socio-economic variables. These include social class, income level, main income source, employment status, whether householder ever had a paid job, number of dependent children, housing tenure and whether the household claims fuel subsidies.

Social Class

The social class of the household, based on the broad category of normal employment of the head of the household, is a key socio-economic variable. In Ireland a standard six-point socio-economic scale is generally used in socio-economic analyses conducted for policy research. Group A consists of professional occupations, while group B contains those employed in managerial and technical positions. C1 refers to skilled non-manual employment, while C2 denotes households whose head works as a skilled manual labour. Group D is comprised of partly skilled persons, while group E encompasses all remaining (unskilled) households.

A very strong relationship is found between the incidence of fuel poverty and social class. As might be expected, the more affluent groups (A, B, C1, C2) have lower levels of fuel poverty than the less well-off groups (D and E); a remarkable one-in-three households in social group E declare fuel poverty (Table 4.9). The relationship between fuel poverty and social class in absolute numbers is characterised by an inverted-U shape, with the highest overall levels found in the most common social groups (i.e. C1, C2 and D); these three groups account for 161,000 fuel-poor households, or three-quarters of all fuel poverty in Ireland. An interesting finding relates to the severity of fuel poverty by social class. Thirty-seven per cent of the fuel-poor in group C1 are demonstrating chronic levels of fuel poverty, compared with 0% in group A. This indicates that more affluent households are having only occasional difficulties in heating their home, and this is perhaps more owing to their dwelling characteristics (larger, less energy efficient) than insufficient income. Despite these plausible results, one anomaly relates to the finding that 15.9% of those in social class A express some intermittent problems with fuel poverty. This may be due to a housing effect; may high-income individuals may inhabit large, older, energy inefficient homes which would require very high heating bills to heat adequately.

Table 4.9 Ability to heat the home adequately by social class (% households)

	N	Some difficulties	Usually not	Never	Total % fuel-poor
A	44	15.9	0	0	15.9
B	92	6.5	1.1	2.2	9.8
C1	412	8.0	2.7	1.9	12.6
C2	356	10.4	1.7	2.8	14.9
D	239	19.2	2.9	0.8	23.0
E	96	30.2	3.1	1.0	34.6

Main Income Source

Dependence on unemployment assistance, state pension or other social transfers implies that such households live on a modest level of household income. Such income may preclude households from heating their home adequately in a direct sense through not being able to afford fuel bills, or indirectly through being unable to invest in household energy-efficiency improvements. The survey found that those whose main income source is social welfare are almost three times more likely to be fuel-poor than those whose income source is paid employment (Table 4.10). Interestingly, chronic and intermittent sufferers of fuel poverty are proportionately similar in both groups, with just over a quarter of total fuel-poor households in each category falling into long-term fuel poverty.

Table 4.10 Ability to heat the home adequately by main income source (% households

	Wages	Social Welfare
N	1,129	307
Some difficulties (%)	8.9	25.1
Usually not (%)	1.5	5.2
Never (%)	1.8	3.3
Total % fuel-poor	12.2	33.6

Employment Status

When the data are disaggregated by employment status, the highest incidence of fuel poverty is found among the long-term ill and disabled, where a remarkable 44.8% are fuel-poor (Table 4.11).[5] The unemployed also fare badly, with some 30.5% declaring fuel poverty. It has been shown by DEFRA and DTI (2001) that the retired are a major risk group in the UK. This study finds the same for Ireland. An overall estimate of 22% is found for the retired, however elderly females appear to suffer significantly higher than males from fuel poverty, with 28.1%

---

[5] Note that this is based on a relatively small sample size.

affected compared with 19.2% of retired males. In addition, female pensioners have among the highest levels of chronic fuel poverty reported in this study; almost 44% of retired female fuel-poor households are suffering from persistent fuel poverty, compared with 28% for males. In absolute numbers, those working at home are demonstrating striking levels of fuel poverty; some 56,000 such households are fuel-poor (the highest group in this section), followed by full-time employees (53,000). Conversely, the self-employed appear to have far less difficulty heating their homes adequately. Regarding the severity of experience, some 42% of those fuel-poor households whose employment status is 'Student / Government Training / Government Employment Scheme' are demonstrating long-term fuel poverty, compared with just 13% of fuel-poor householders working part-time. The chronic fuel poverty experienced among students is likely to be a result of the large numbers who live away from home on modest incomes and, consequently, living in poor housing conditions.

Table 4.11 Ability to heat the home adequately by employment status (% households)

	N	Some difficulties	Usually not	Never	Total % fuel-poor
Self-employed	131	7.6	2.3	0.8	10.7
Employee (Full-time)	554	7.0	1.6	2.7	11.4
Employee (Part-time)	112	17.9	0.9	1.8	20.5
Unemployed*	59	22.0	5.1	3.4	30.5
At home	302	18.5	2.3	1.0	21.9
Ill/Disabled	29	34.5	6.9	3.4	44.8
Retired (Male)	125	15.2	2.4	1.6	19.2
Retired (Female)	57	15.8	3.5	8.8	28.1
Student/Govt. training/ Govt. work scheme	100	11.0	7.0	1.0	19.0
Other	10	-	-	-	-

* Includes those not working but seeking to work, those seeking work for 1st time, and those not working and not seeking work.

Whether Householder Ever Had Paid Employment

It is worthwhile to cross-tabulate the results by whether the householder has ever been in paid employment. Households which have not had a paid position (almost 10% of the sample) report substantially higher levels of both chronic and intermittent fuel poverty, with some 24.2% overall (28,000 households) in fuel poverty. This compares with 16.7% for those who have had paid employment, as can be seen in Table 4.12. Chronic fuel poverty is highest among those who have not had paid employment; persistent fuel-poor households account for 36.4% of all fuel-poverty sufferers in this ('No') group, compared with 25.1% among the 'Yes' category.

Table 4.12 Ability to heat the home adequately by whether householder ever had paid work

	Yes	No
N	1,333	149
Some difficulties (%)	12.5	15.4
Usually not (%)	2.2	5.4
Never (%)	2.0	3.4
Total % fuel-poor	16.7	24.2

Household Income

There is a very strong correlation between fuel poverty and income, as would be expected given that income is an important component in the definition of fuel poverty. In addition, both the incidence and actual numbers affected by fuel poverty show a very similar linear relationship with gross household income.

Table 4.13 Ability to heat the home adequately by income (% households)

Gross household income p.a. (2001)	N	Some difficulties	Usually not	Never	Total % fuel-poor
<€8,888	162	28.4	6.2	4.9	39.5
€8,889-12,696	141	27.7	2.1	1.4	31.2
€12,697-19,045	185	11.4	3.2	2.2	16.8
€19,046-25,393	229	11.8	0.9	1.7	14.4
€25,395-33,012	183	7.7	3.3	2.2	13.1
€33,013-38,091	107	4.7	0	3.7	8.4
>€38,092	131	4.6	0.8	0.8	6.1

Some two-in-five households under €8,888 are suffering fuel poverty.[6] Those on incomes between €8,889 and €12,696 also report very high levels of fuel poverty, with some 31.2% in fuel poverty. Conversely, those on relatively high incomes suffer far less from fuel poverty. Those on over €38,092 demonstrate an incidence of just 6.1%, with virtually no-one (0.8%) in this high-income group declaring chronic fuel poverty (Table 4.13).

The significant proportion of high-income households suffering some level of fuel poverty is likely to be, in no small part, a product of thermal inefficiency. Such households are either making an informed choice not to heat their homes comfortably (their homes may be very large and expensive to heat), or

---

[6] €1=£0.70=$1.18 (November 2003).

they may be very energy inefficient making it difficult for even high-income households to afford adequate warmth.

Number of Dependent Children

This section is notable, as the incidence of fuel poverty does not appear to be very substantially affected by the number of dependent children; a weak U-relationship is reported. However, the actual number of households affected varies very significantly. The largest group of fuel-poverty sufferers are those without any dependent children; some 143,000 households are either persistent or intermittently fuel poor, by far the largest group suffering (Table 4.14). An interesting result is found with severity of experience. A very high incidence of persistent fuel poverty is found among those households with four or more dependent children; some 55.5% of fuel-poor households with four or more dependants fall into the chronic group of fuel-poverty sufferers. This indicates that many large families are finding it difficult to adequately heat their home over time. Such a result is worrying given that the health effects of cold and damp exposure are particularly acute among children.

Table 4.14 Ability to heat the home adequately by number of dependent children (% households)

	N	Some difficulties	Usually not	Never	Total % fuel-poor
0	918	13.2	2.9	2.5	18.6
1	122	10.7	1.6	3.3	15.6
2	231	12.1	1.7	0.9	14.7
3	158	13.3	1.9	0	15.2
4+	60	9.3	2.3	9.3	20.9

Housing Tenure

Housing tenure is an important dynamic of fuel poverty, as has been noted by Whyley and Callender (1997). This is because it gives households varying levels of control over their home, heating systems and their energy consumption. Owner-occupiers may be considered as fully autonomous, while tenants may be more limited as regards what they feel they can afford to do to improve their housing or even what they are authorised to do to improve their housing (Clinch and Healy, 2000a).

Remarkably high rates of fuel poverty are calculated for tenant households, but especially those who rent from a Local Authority, where one-in-three such households are unable to adequately heat their home, compared with just one-in-ten owner-occupier (mortgaged) households (Table 4.15). Privately rented households also suffer highly from fuel poverty, with an incidence of 24.5%. In absolute numbers, the highest levels of fuel poverty are found among those who own outright their home (85,000 households), followed by local authority-rented households (57,000). In terms of the severity of fuel poverty, some 44.1% of privately rented fuel-poor households suffer chronic fuel poverty, compared with 19.3% of the fuel-poor renting from local authorities.

Table 4.15 Ability to heat the home adequately by housing tenure (% households)

	N	Some difficulties	Usually not	Never	Total % fuel-poor
Owned outright	618	11.5	2.1	2.6	16.2
Owned (mortgage)	533	7.9	0.9	1.9	10.7
Rent (private)	102	13.7	7.8	2.9	24.5
Rent (local authority)	198	27.3	5.1	1.5	33.8
Other	20	-	-	-	-

Fuel Allowance

It is interesting and useful to examine whether (low-income) households in receipt of fuel subsidies (known as the 'Fuel Allowance' in Ireland) report higher or lower levels of fuel poverty than those households not claiming such social transfers. This is because a result showing a higher incidence of fuel poverty among claimants of the subsidy would imply that such an allowance is not sufficient in meeting home-heating costs; this would most likely be due to the interaction between poor thermal efficiency and low household income. Conversely, a lower incidence among claiming households could imply that the allowance is at least partly successful in assisting fuel-poor homes in meeting their domestic fuel costs. However, it could also imply that the allowance should be widened. An insignificant result (where the fuel subsidy is shown to be ineffective in changing the rate of fuel poverty) could imply that the fuel allowance needs to be radically re-designed in structure.

The results demonstrate the first case scenario. Households in receipt of fuel allowances in the past 12 months report an incidence of fuel poverty of 37.9%, compared to just 13.8% among non-claiming households (Table 4.16). However, it should be noted that proportionately less fuel-poor households claiming the fuel allowance (19.7%) are suffering from persistent fuel poverty, while 29.0% of non-claiming households suffering fuel poverty are chronic sufferers. It is perhaps logical to surmise that the fuel allowance, while ineffective in reducing the overall level of fuel poverty in Ireland, does reduce the severity of experience of fuel poverty among the low-income households it covers.

Table 4.16 Ability to heat the home adequately and households claiming fuel subsidies (% households)

	N	Some difficulties	Usually not	Never	Total % fuel-poor
Fuel allowance	195	30.8	4.1	3.1	37.9
No fuel allowance	1262	9.8	2.1	1.9	13.8

Fuel Poverty and Housing Conditions

There are a number of adverse housing conditions often believed to be associated with fuel poverty. These are generally manifestations of the energy inefficiency of the dwelling and examples include the presence of damp or mould on walls, floors and ceilings and the presence of condensation on walls, ceilings and windows. The former housing condition has been shown to be strongly linked with an increased risk of ill health, mainly through exacerbations of respiratory and cardiovascular disorders such as asthma and bronchitis. Such effects are most acute in the very young and the old (Collins, 1986). This part of the study examines whether there is a relationship between the incidence of fuel poverty and the presence of such unwanted housing conditions in Ireland.

Damp

The survey results show that 10.3% of households suffer from the presence of damp in Ireland. Of these households, 48.1% are reporting some level of fuel poverty, although three-quarters (74.2%) of these households are intermittent sufferers. Conversely, households without problems of damp report much lower levels of hardship, with just 13.8% suffering fuel poverty. In physical numbers, some 63,000 damp households are fuel-poor.

The relationship between damp housing conditions and fuel poverty can be further examined on a room-by-room basis. Care should be taken in drawing strong conclusions from these results, as they are based on relatively small sample sizes of between 24 and 59 households. However, it appears that the strongest relationship with fuel poverty is found among households reporting mould in bedrooms, especially the master bedroom; this would appear to be logical, bearing in mind that one-third of our lives are spent in this room on average. An aggregate incidence of fuel poverty of 66.7% is calculated; 29.2% of these households suffer chronic fuel poverty. There is also a high association found among households with damp bathrooms, with 51.9% of such households displaying fuel poverty, 35.6% of which are chronically fuel-poor (Table 4.17).

Table 4.17 Ability to heat the home adequately and presence of damp (% households)

	N	Some difficulties	Usually not	Never	Total % fuel-poor
Damp house	154	35.7	7.8	4.5	48.1
No damp	1,331	10.0	1.9	2.0	13.8
Damp living room	24	16.7	8.3	12.5	37.5
Damp kitchen	59	30.5	6.8	8.5	45.2
Damp master bedroom	48	47.9	14.6	4.2	66.7
Damp 2nd bedroom	52	48.1	11.5	5.8	65.4
Damp 3rd bedroom	36	36.1	11.1	8.3	55.6
Damp bathroom	54	33.3	11.1	7.4	51.9

Condensation

The study finds that condensation, like damp, is strongly associated with fuel poverty. Some 39.2% of households in Ireland with condensation problems report some level of fuel poverty, compared to just 13.5% of households without condensation. About 39.3% of fuel-poor households with condensation are persistent sufferers of fuel poverty (Table 4.18). Overall, almost a quarter (23.0%) of Irish households suffer from condensation problems which implies substantial public-health concerns.

Table 4.18 Ability to heat the home adequately and presence of condensation (% households)

	N	Some difficulties	Usually not	Never	Total % fuel-poor
Condensation	341	23.8	3.8	2.6	39.2
No condensation	1,143	9.3	2.1	2.1	13.5

Overall, this study finds that the fuel-poor are more likely to live in damp homes and in housing with condensation. Such findings are of obvious concern for those responsible for housing policy and public health. The next section subjects the results of this chapter to more sophisticated statistical procedures.

Multivariate Analysis of Fuel Poverty

To further test the reliability of the results presented heretofore in this chapter, this section outlines the results of a multivariate-regression exercise performed on the dataset. A Probit regression analysis is used to examine those factors that influence the probability of being fuel poor. A number of indicators of fuel poverty are regressed against socio-economic and other characteristics of households. A Probit analysis allows us to examine the household characteristics that are significantly associated with each indicator of fuel poverty. The added value over the bivariate analysis presented in the cross tabulations is that we can examine the effect of each variable holding all else equal.

Table 4.19 Predictive power of the Probit regression for ability to heat home adequately (%)

	Predicted	Observed
Yes, without any problems	82.8	82.7
Usually with occasional difficulty	12.5	12.6
Usually not,	2.3	2.4
No, not at all	2.2	2.2

The predictive power of the ordered-Probit model for the household's ability to heat the home adequately is good as seen in Table 4.19. The results of the model suggest that, *inter alia*, being better educated, in a newer house and being a couple are significantly associated with being able to heat the home adequately (Appendix 4.1). Being a tenant and living in rent-free accommodation are significantly associated with being much less likely to be able to heat the home without any problems. The marginal effects suggest that housing tenure is most strongly related to the inability to heat the home adequately.

Reasons for Non-Investment

Energy efficiency in the home makes good economic sense. The costs and benefits of improving the thermal efficiency of the Irish dwelling stock to the latest building regulations have been quantified by Clinch and Healy (2001), who report a net social benefit of some €3.1 billion. The private and external benefits (energy, environmental, health and comfort) outweighed the labour and materials' costs of the programmes by 3:1, while the private benefits to the individual household in terms of reduced fuel bills and increased health and comfort are very substantial (a private benefit-cost ratio of 1.7:1 is calculated). If the benefits of domestic energy-efficiency are so great, why then do households not invest in retrofitting these measures into their homes. This question was asked to the survey respondents who are defined energy inefficient (i.e. they lack basic household energy-saving measures), and the results are illustrated in Figure 4.1.

Clinch and Healy (2000b) hypothesised the reasons for market failure in domestic energy efficiency in Ireland. This study identifies these reasons empirically by asking respondents why they do not invest in energy-saving measures when the benefits clearly outweigh the costs.

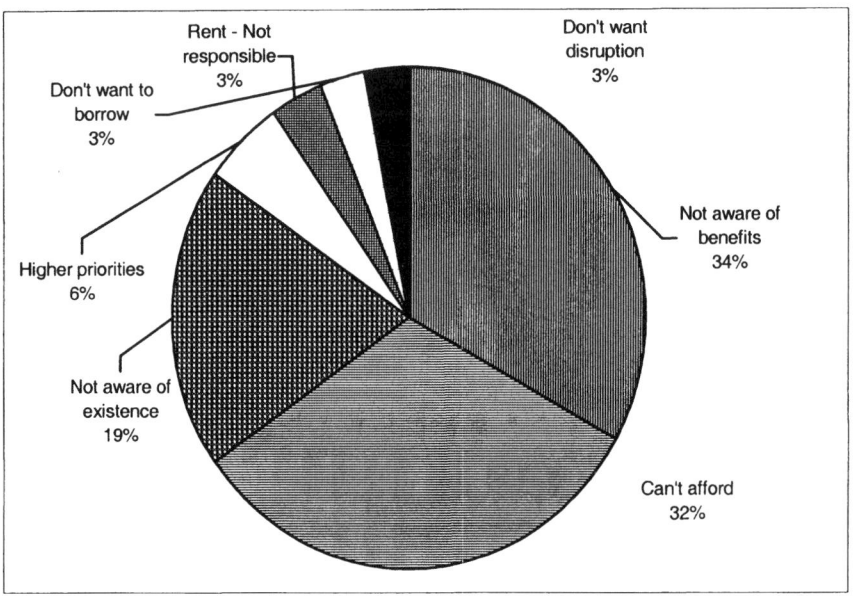

Figure 4.1 Reasons for non-investment in energy-saving measures in Ireland

Information Gap

One of the principal reasons for financially viable energy-conservation measures not being taken up is the lack of knowledge on the part of householders of the opportunities for saving on fuel bills. This information gap is likely to be greater in low-income households where the benefits would be greatest. In addition, an information asymmetry between buyers and sellers of energy-efficiency measures may occur, leading to adverse selection of such technology.[7] Lack of information is seen as a key reason for market failure in the UK according to Williams and Ross (1980) and Carlsmith et al. (1990), and this study concurs strongly with British policy analysis of market failure in domestic energy conservation.

The results of the UII national household survey demonstrate a large 'information gap' in the market for domestic energy efficiency. Some 32.3 per cent

---

[7] See Smith (1992).

of energy inefficient households are not *aware* of the benefits of energy-saving measures, while a further 19 per cent did not know of their *existence*. This implies that over half of energy inefficient households in Ireland (amounting to approximately 120,000 homes) are unaware of the benefits of retrofitting these measures into their homes, despite on-going information and awareness campaigns funded by the Irish government; as such, this appears to be, by far, the strongest impediment to the diffusion of energy efficiency in the residential sector. This result has clear implications for policymakers.

In addition, if the housing market worked effectively, the monetary value of the energy-efficiency measures would be reflected in the re-sale value of the house. However, if the public is lacking in knowledge as regards the benefits of the measures, this will not happen. Therefore, if individuals are likely to move house in the meantime, they may not be willing to make an investment with a long payback period.

Socio-economic Considerations

The least energy efficient households are more likely to be lower income households (Clinch and Healy, 1999a). Such households are much less likely to have available funds and, thus, are most likely to have to resort to a loan. They are less likely to be in the position of accessing credit (particularly at the market rate of interest)[8] and they are more likely to have more pressing alternative uses for any extra funds. They may, additionally, have an aversion to borrowing funds, as has been reported by Salvage (1992). It has also been shown that low-income households tend to have higher discount rates, i.e. they exhibit myopic tendencies whereby they place a greater value on income now as opposed to in the future, partly resulting from the higher degree of uncertainty about the future stemming from their financial instability. Therefore, *ceteris paribus*, such households are unlikely to invest in something that might not pay for itself for over 30 years.

These financial constraints are found to be significant barriers in the market for domestic energy efficiency. Some 31.6 per cent of respondents (75,000 households) reported an inability to pay for these measures, while a further 5.5 per cent (13,000 households) reported more pressing priorities for expenditure. In addition, borrowing constraints were identified by another 3 per cent of householders (7,000 households). Therefore, over 40 per cent of energy inefficient households blame income constraints as barriers to improving their home's thermal efficiency.

Property Rights Failure

Some of the least energy efficient houses in the UK and Ireland are tenant-occupied (Brechling and Smith, 1992; Healy, 2002a). Tenants may feel that they are not responsible for undertaking investments in energy efficiency or authorised to do so. Indeed, it is not financially sound for a tenant to invest if they expect to

---

[8] See Weber (1990) for more on this issue.

move out in the short to medium term. Likewise, landlords may feel that the benefits to them of such investment may not be recouped if they are unable to raise rents. Also, if investment does take place in a multi-occupancy dwelling, 'free-rider' incentives may exist in relation to the financing of the public good (Smith, 1992).

The results of the UII national household survey of Ireland indicate that some 8,000 households (3.4 per cent) did not feel responsible to undertake the retrofit because they rented the dwelling they occupied and felt it was the landlord's responsibility.

Transactions' Costs

Another potential 'blockage' in the market for energy-efficiency measures is that of the fixed costs of learning about, and administering, energy-conservation measures. Examples of transactions' costs include the time householders must spend to learn about the various options, locate a suitable installer and oversee the work. Some householders may also be concerned about the appropriate techniques and the quality of the workmanship, as well as the attendant disruption of installing these measures. Such costs are not reflected in Cost-Benefit Analysis and, therefore, the full costs of retrofitting households with energy-conservation measures may be significantly higher to the individual than is suggested by the figures. The amplitude of these transactions' costs may overwhelm the potential pay-off of such an effort, acting as a performance-inhibiting 'wedge' which prevents the implementation of cost-effective energy-conservation measures in the home. These transactions' costs are difficult to measure, but have been seen as potential factors in explaining the slow take-up of financially viable measures in the UK,[9] especially in the domestic sector. The results of this study do not corroborate the hypothesis that transactions' costs act as a major impediment to the diffusion of energy-saving measures in Irish households. In fact, just 3 per cent of energy inefficient households (about 7,000 homes) blame such costs as the major reason for not installing these measures.

Discussion

This chapter employed new survey data on home heating to assess, for the first time, the severity of fuel poverty across Ireland and to identify the social groups most vulnerable. It was demonstrated that, while the penetration of lagging jackets, double glazing and central heating have improved substantially over the past five years, the Irish housing stock remains considerably under-protected from the outdoor environment, leaving the vulnerable in society open to fuel poverty and increased risk of ill health. A national estimate of fuel poverty of 17.4% (or 226,000 households) is produced for Ireland. This estimate is close to the estimate

---

[9] This slow diffusion of apparently cost-effective energy-conservation technologies across households has been denoted the 'energy paradox' by Jaffe and Stavins (1994a, 1994b).

of 16.4% produced by DEFRA and DTI (2001) for the UK (which includes housing benefits in the calculation). This research also identified sufferers of fuel poverty by severity. It is estimated that 27% of fuel-poor households (4.7% of the total housing stock) is suffering from chronic fuel poverty, where householders are caught in a persistent fuel-poverty trap, constantly unable to adequately heat the home. It is calculated that 12.7% of all households suffer from intermittent levels of fuel poverty, where householders are occasionally unable to heat the home adequately.

The highest incidence of fuel poverty is found among the long-term ill and disabled, where 44.8% of such households (11,000 in absolute numbers) are demonstrating fuel poverty. However, this result is based on a relatively small sample of households and should be treated with care. Lone-parent households are identifiable as the second-highest group of fuel-poverty sufferers in Ireland, with over two-fifths (40.2%) declaring an inability to heat the home adequately (29,000 households). Income is closely associated with fuel poverty, and the results of this study bear out this tenet; 35.6% of households under €12,700 are suffering (some 121,000 households). A similarly strong result is found with social class, where 34.6% of those in the lowest group (E) are reporting fuel poverty (34,000 households). Other groups with high incidences include: local-authority tenants (33.8% or 57,000 households), the unemployed (30.5% or 16,000 households), one-person occupied households (28.4% or 51,000 households), those separated, divorced or widowed (28.2% or 37,000 households), lone female pensioners (28.1% or 11,000 households) and those who completed their education at primary level (25.6% or 47,000 households). Chronic fuel poverty is proportionately highest among households with four or more dependent children, where 55.5% of fuel poverty appears to be persistent, followed by 44.1% of fuel-poor private tenants and 42% of students living in fuel poverty. Because the risk groups are all low-income households, it is likely that income subsidisation combined with investment in energy-saving measures would be required to lift them out of a fuel-poverty trap.

The results of the survey demonstrated a strong association between fuel poverty and household condensation and an even stronger association with household damp, where the incidence of fuel poverty is almost four times that found among households without damp spores. The study also indicated that the fuel allowance is a necessary but insufficient measure in tackling fuel poverty. Households claiming the fuel subsidy report an incidence of fuel poverty some three times higher than non-claiming households, however the fuel allowance appears to impact positively on the severity of experience, significantly reducing the proportion of chronic fuel-poor households.

The study also presented the results of the survey regarding the reasons why energy-saving measures are not adopted by households. The findings indicate a large 'information gap' in the market of domestic energy-efficiency measures, with over a half of households either unaware of the existence or unaware of the benefits of energy-saving measures in the home. In addition, over a third of households identified financial constraints to retrofitting, with only a very small proportion blaming transactions' costs. Taken as a whole, these results argue for government intervention to rectify the market failure. The policy implications of this chapter are explored in full in Chapter 10.

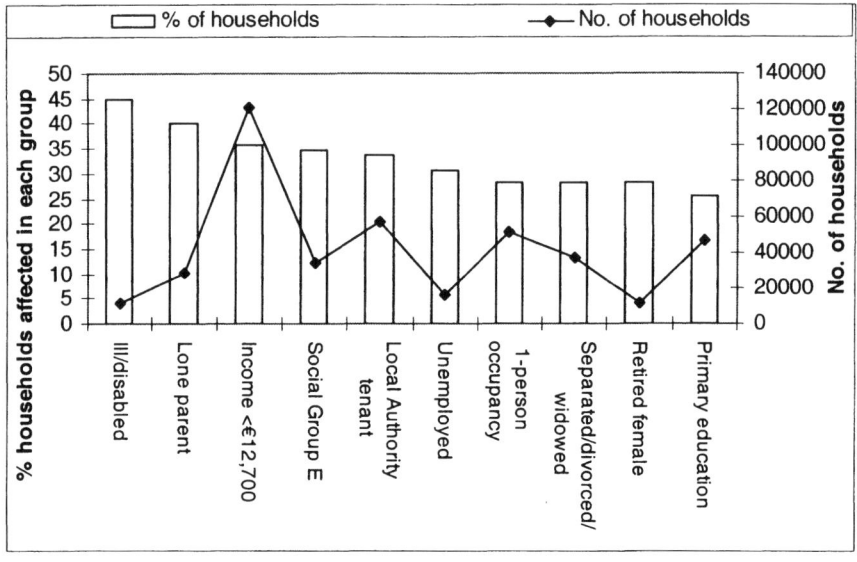

Figure 4.2 Fuel poverty in Ireland: 10 risk groups

Chapter 5

Housing Deprivation and Self-Reported Health in the EU

Introduction

Chapters 2, 3 and 4 have presented, in some detail, the cross-country results of housing deprivation and fuel poverty. This chapter builds on this work (especially that presented in Chapter 2) by analysing the impact of housing deprivation on fuel poverty using European data.

Studies analysing the determinants of inequalities in health have, over the past decade or so, drifted toward a consensus regarding the major explanations for the 'health divide' apparent in modern industrial societies. While cultural and genetic explanations have some relevance, most health inequalities are explained by what Townsend has called 'materialist' or 'structural' explanations (Townsend, 1992). In addition, class inequalities in health have become understood in terms of aspects of the socio-economic environment, such as work conditions, household overcrowding and smoking rates. Townsend's seminal study concluded that much of the inequality can only be understood in terms of the diffuse consequences of the class structure in terms of poverty, work conditions and deprivation in its various forms (Townsend, 1979).

Poor people have been shown to suffer from higher levels of ill-health than the more affluent for a variety of 'materialist' reasons. The less well-off may lack the necessary resources for achieving a healthy lifestyle; in particular, this can entail inadequate, unhealthy diets (Forsyth et al., 1994). Furthermore, they may also expose themselves to health risks working in more dangerous, risky environments where significant health hazards exist (Waitzkin, 1997). The poor may expose themselves to higher levels of lifestyle risk factors strongly associated with ill health, such as high smoking rates (Mannino et al., 2001). It has been shown by McMahon (1993) and others that people on low incomes often lose out in the provision of healthcare support. In addition, they may live in poor housing conditions because of low household income and expose themselves to chronic cold and damp conditions (Clinch and Healy, 1999b). All of the above factors play key roles in explaining the modern-day 'health divide'.

It is now over 20 years since Peter Townsend's path-breaking research on UK poverty and inequality, which stressed the importance of the home (i.e. housing conditions) as a potential causative factor in poor health. Within a decade or so, Brenda Boardman stated that large proportions of households in the UK were so deprived in their housing conditions, and so inadequately protected from the cold, that significant numbers were dying each year during Britain's winter months because they were fuel-poor. Since then, the UK government has recognised fuel poverty as a *bona fide* social problem. However, there is still a large degree of uncertainty regarding the relative importance of fuel poverty and poor housing conditions on human health. There is a growing body of research testing this health linkage. Recent research testing the association between ill health and poor housing conditions, especially in relation to inadequately heated, damp housing, has been demonstrating strong adverse health implications (Marsh *et al.*, 1999; Rudge and Nicol, 2000), however strong causal relationships are difficult to identify because of confounding variables. Furthermore, almost all of this research is conducted in the UK – where most fuel-poverty research has occurred – and very little analysis has assessed the relationship between housing conditions and health in the rest of Europe, especially in Mediterranean countries. In addition, there is a dearth of research that identifies the health effects of housing dissatisfaction and housing affordability. This research deficit has occurred, not because of a lack of interest in the area, but because of some major logistical reasons, most obviously the lack of comparable cross-country data.

In addition to public-health implications of poor housing conditions and fuel poverty, there is a strong environmental-policy perspective, as has been outlined previously. This chapter tests the relationship between health and a number of consensual social indicators of housing deprivation across 14 countries in Europe using longitudinal data from the first pan-European survey on social conditions, the European Community Household Panel. The data cover the four-year period 1994-97, and the following indicators are analysed:

- Housing conditions
- Housing satisfaction
- Housing affordability
- Fuel poverty

Causal relationships are not easy to identify. This study employs a methodology which is similar to much public-health research insofar as it attempts to examine associations found among data. It should be noted that firm conclusions and causal relationships are not likely to be drawn from such an approach.

The study's European study frame is important, not just because there is so little cross-European research on health and housing, but because there is a growing concern that large proportions of Europeans are living in sub-standard housing, with households considerably under-protected from the outdoor environment (Healy, 2001a). The large variations in seasonal mortality in Europe have been associated with inadequately heated homes in Ireland, the UK and other parts of Europe (Clinch and Healy, 2000c; Eurowinter Group, 1997), so there is a strong need to assess whether poor housing is responsible for other adverse health outcomes throughout the Member States. The next section describes the European context in more detail and there follows a subsequent section outlining the methodology for the study, before the cross-country results are presented. In

Housing: the European Context

Housing standards, especially those pertaining to energy efficiency, vary considerably across Europe, as has been shown by Healy (2001a). Of course, certain countries prioritise thermal measures in the design and construction of new housing, as it is essential protection to combat the relatively severe winters experienced in these colder climates where winter temperatures are often below freezing (Clinch and Healy, 1999a). Nonetheless, Ireland and the UK have the highest rates of seasonal mortality in northern Europe, and it has been shown that such mortality rates result, in no small part, from the inadequately protected, thermally inefficient housing stocks in these countries (Curwen, 1991; Eurowinter Group, 1997). There are also studies showing strong associations between inadequately heated homes and increased rates of morbidity; higher incidences of various cardiovascular and respiratory diseases have been associated with cold exposure from within the home (Collins, 1986; Evans et al., 2000). Thus, when temperatures fall during a typical British or Irish winter, households need to increase their expenditure on fuel considerably to heat their home adequately, owing to the poor level of heat retention in their dwellings. The problem of fuel poverty occurs, therefore, when a household does not have the adequate financial resources to meet these winter home-heating costs, and because the dwelling's heating system and insulation levels prove to be inadequate for achieving affordable household warmth. A key aim of this chapter is to strengthen the link between poor housing standards, fuel poverty and self-reported poor health.[1]

Data on housing conditions and energy-efficiency levels in southern Europe are notoriously difficult to obtain, and their reliability is often questionable. However, the analysis of general housing conditions presented in Chapter 4 indicated that high levels of energy efficiency in southern-European housing are not prioritised in building regulations. It is an often-overlooked fact that many parts of southern Europe also face cold winter temperatures, yet their housing stocks appear to be poorly protected from the cold, and they are also the poorest countries in Europe using measures such as income poverty and inequality as well as macroeconomic indicators like GDP per capita.[2] Despite this, there has been virtually no published research on housing conditions, fuel poverty and health in southern Europe, despite a growing body of research on this area in the UK and Ireland. This study attempts to rectify this research deficit by testing the hypothesis that households reporting selected indicators of housing deprivation in countries that exhibit poor housing conditions and high levels of fuel poverty (and domestic

---

[1] There is some ongoing research in this area in the UK, but very little elsewhere in Europe. Rudge and Nicol (2000) present an overview of current British research in this area.
[2] Italy being the exception here, with GDP per capita considerably above EU-average.

energy inefficiency) also realise the highest proportionate variations in the incidence of self-reported poor health.

In addition to fuel poverty and energy inefficiency, there are other housing conditions which are thought to provoke adverse health outcomes. These conditions include overcrowding, unaffordable housing costs and general housing dissatisfaction. In relation to the former, there is a long-standing debate about the relative importance of household crowding, and the literature showing the relationship between crowding and ill health reports somewhat mixed results. Chapter 4 presented results which demonstrated there are significant variations in the levels of overcrowding across EU-14, with the highest incidences in southern countries, so this study aims to shed some light on the relationship between household crowding and health using pan-European data. A very interesting study by Konadu-Agyemang (2001) identifies lack of affordable housing as a key factor in the "dismal housing situation" facing prospective home-owners in Accra and he demonstrates strong links between affordability and satisfaction with housing. The lack of affordable housing, and the associated high housing costs, is a considerable problem in Europe, as Chapter 4 demonstrated. The link between housing satisfaction and happiness is well-researched and highly correlated; indeed, it has been shown by Barresi *et al.* (1984), amongst others, to be the key factor influencing happiness, especially in older populations. In light of these facts, the final part of the analysis on housing and health in this chapter tests for a relationship between housing satisfaction and health status across Europe.

Health and Housing Conditions

Households in EU-14 are asked about various aspects of their housing conditions in the ECHP survey each year. The indicators selected for this analysis pertain to the energy-efficiency of the dwelling and to overcrowding. The questions include asking householders about physical attributes, such as whether they have leaky roofs, damp walls, ceilings or floors or rotten windows. In this section, the incidence of self-reported poor health (households reporting either "very bad" or "bad" health status) is analysed against these self-reported housing conditions to assess if such households are displaying significantly higher rates of poor health. As should always be noted with self-reported data, there are some important caveats, especially in relation to the margin of error which can be attributed to self-reporting. However, the very large sample size of the ECHP dataset acts, somewhat, as a compensating factor regarding the potential margin of error normally expected with self-reporting.[3]

---

[3] Most countries are 'over-sampled' in the ECHP. This is because households are followed and re-surveyed year after year and, as would be expected, the number of respondent households declines each year (some move address, others are unavailable for interview, etc.). To illustrate the level of sampling, over 4,000 households were interviewed successfully in Ireland for 1994. This is almost three times the number necessary to be statistically significant.

Inability to Adequately Heat Home

The analysis begins by analysing the levels of self-reported poor health in those households declaring an inability to adequately heat the home – a key indicator of fuel poverty – against those that do not declare such difficulties. Gordon *et al.* (2000), using data from a recent UK Omnibus Survey, reported that the ability to heat living areas of the home adequately is regarded as essential by 94% of respondents, making this socially perceived necessity the strongest indicator of material deprivation after 'the ability to provide a bed for everyone in the household' (95%). Energy inefficiency in the home can make home-heating such a burden to low-income householders that they cannot afford to heat their home adequately and, as such, live in fuel poverty (Boardman, 1991). The implications on human health of living in fuel-poor conditions are thought to be particularly pernicious and have been shown to be associated with higher incidences of various cardiovascular and respiratory disorders (Collins, 1986). In addition, chronic, long-term exposure to cold and damp in the home has been associated with premature mortality (Gemmell *et al.*, 2000). This analysis tests the hypothesis that fuel poverty and inadequately heated homes results in higher incidences of poor health status.

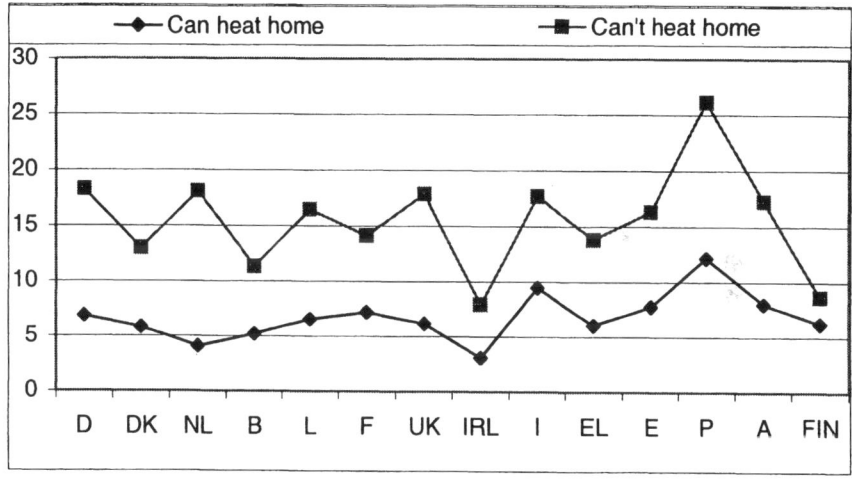

Figure 5.1 Poor health and ability to heat home adequately (mean % of households, 1994-97)

The results show a clear relationship between fuel poverty and self-reported poor health status (Figure 5.1). The proportion of households demonstrating bad health is significantly higher among fuel-poor households (those that cannot adequately heat their home) for each of the 14 Member States under analysis. Across EU-14, 6.8% of households that can afford to adequately heat their home report ill health

(95% confidence intervals (CI) of 5.5% and 8.1%), compared with 15.5% of those households unable to adequately heat their homes (CI=12.9, 18.2, α=0.05). T-tests demonstrate that the results are highly significant (P<0.001). A strong, statistically significant correlation is found between both sets of self-reported health data across EU-14 (R=0.75, P=0.002). This indicates that the data are 'behaving'; countries with high levels of poor health status among households that *can* afford to heat their homes are consistently reporting higher levels of poor health among households that *cannot* afford to heat the home, and *vice versa*, as theory would predict. These results, based on subjective data, appear to corroborate the results of Collins (1986) and others in showing an association between sub-optimally heated homes and decreased health status, though they point to no specific causality.

The most notable differences occur in the Netherlands (where 14.3% more households declare poor health if fuel-poor), Portugal (14.0% variation), the UK (11.7%) and Germany (11.4%). The highest levels of poor health status are reported in Portugal, where 26.2% of fuel-poor households are declaring either "very poor" or "poor" health. However, perhaps the most important findings relate to the *proportionate* variations reported in health status among households declaring the deprivation indicator. Proportionately, the largest variations in health status occur in the Netherlands, where fuel-poor households are some 3.4 times more likely to declare poor health status than other households, followed by the UK (1.9 times higher incidence among fuel-poor households). Conversely, proportionate increases in levels of poor health are far smaller for Finnish households.

Adequate Heating Facilities

Households without adequate heating systems find it difficult to heat their homes adequately at an affordable cost and, as such, may be living in fuel poverty. The relationship between poor health and fuel poverty was shown to be strong in a previous section, so it is useful to assess the relative importance of the heating system (one of the three crucial components of the fuel-poverty equation) as a determinant of health status.

The results in this analysis demonstrate that households without adequate heating facilities demonstrate substantially and persistently higher levels of poor health across Europe, as might be predicted. An average of 12.0% of households without adequate heating facilities across EU-14 are reporting poor or very poor health status (confidence intervals of 9.1% and 14.9%, α=0.05) compared with 7.4% of households without this adverse housing condition (CI=5.4%, 9.5%, α=0.05). T-tests reject the null hypothesis that these variations are insignificant, and a very strong P-value is found (P<0.001). The correlation between the incidence of poor health among households lacking adequate heating facilities and those households with adequate facilities is very strong (R=0.92, P<0.001), and all countries report higher incidences of poor health among households lacking adequate facilities.

The biggest percentage-point variations in the incidence of poor health status occur in Italy (8.8% difference in the incidence of poor health between households lacking adequate heating facilities and those with adequate heating facilities), Greece and Spain (8.3% respectively). The highest overall level of poor health is found in Portugal, where almost a quarter (24.1%) of households lacking adequate heating facilities are self-declaring poor health. Ireland appears to demonstrate the largest proportionate variations in health status, with an incidence of poor health some 1.5 times higher among households displaying the lack of this socially perceived necessity. The Netherlands and Greece follow closely, with an increased incidence of poor health of approximately 127% and 95% above the incidence among households with adequate facilities.

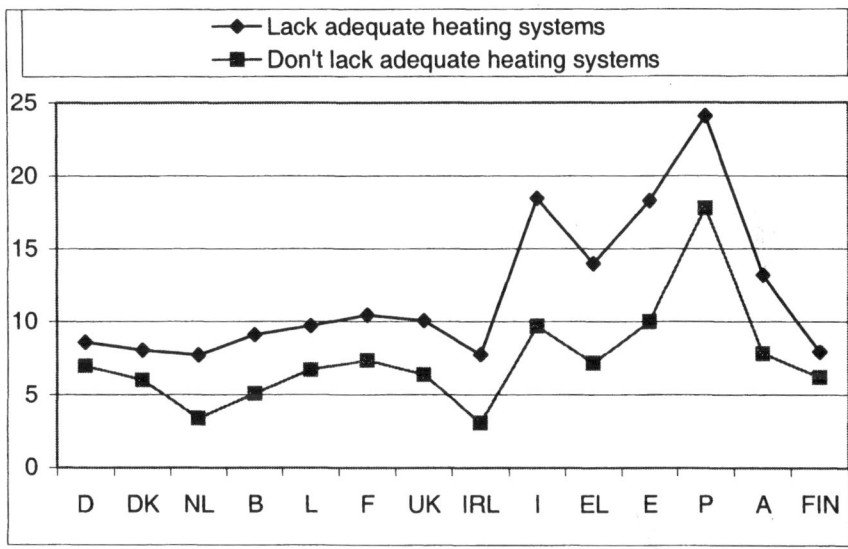

Figure 5.2 Poor health and adequate heating facilities (mean % of households, 1994-97)

Damp

The presence of damp indicates that the dwelling may not be energy efficient. It may also be a manifestation of a continuously unheated or ineffectively heated home. In both cases, it acts as a good objective indicator of fuel poverty, as well as a strong indicator of housing deprivation (Marsh *et al.*, 1999). In fact, Gordon *et al.* (2000) reported that 93% of British households regard a damp-free home as a key socially perceived necessity (third-highest in a list of over 50 items). There is a very substantial epidemiological and public-health literature showing causal relationships between household damp and ill health, and particularly regarding an increased incidence of asthma (Korsgaard, 1983; Salvaggio and Aukrust 1981; Williamson *et al.*, 1997).

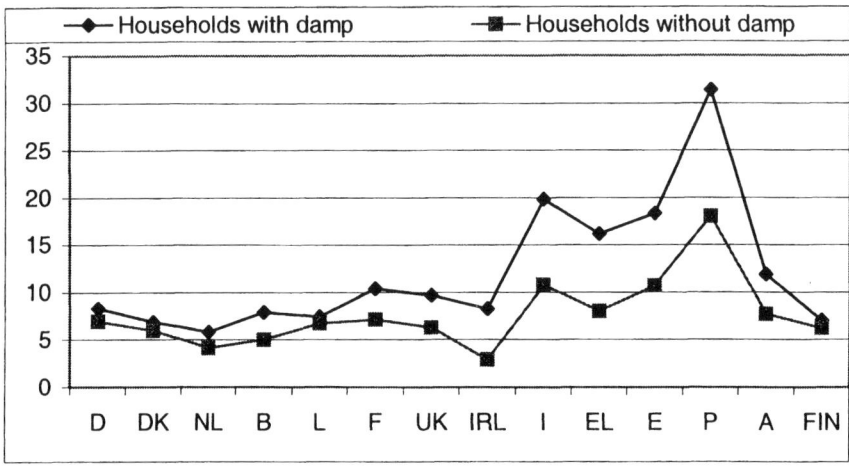

Figure 5.3 Poor health and damp (mean % of households, 1994-97)

Households are assessed to identify if there are any patches of damp on either the walls, floors or foundations in their home. Households are, again, split into those reporting the presence of household damp and poor health status versus those reporting poor health status but no household damp. The results corroborate the vast majority of medical literature in showing a strong relationship between damp housing and poor health status. Across EU-14, a consistently higher incidence of poor health status is reported among households with damp; mean levels of 12.1% (CI=8.0%, 16.2%, α=0.05) and 7.6% (CI=5.5%, 9.8%, α=0.05) are found. T-tests reject the null hypothesis that these variations are not significant (P<0.001), and a correlation coefficient of 0.95 (P<0.001) is found between the increased poor health among households with damp against those households with poor health and no problems of damp.

The largest percentage-point variations are found in Portugal (13.3%) and Italy (9.1%). The highest overall level of poor health is found in damp households in Portugal, where some 31.4% are affected, followed by 19.9% in Italy. The largest proportionate variations in health are found again in Ireland, with an incidence 182% higher among households declaring damp, followed by Greece (102%) and Italy (84%).

Rot

Window frames which have become rotten are not energy efficient and, as such, can be considered a good (objective) indicator of fuel poverty and general housing deprivation. Again, households declaring this housing condition and reporting poor health status are compared against those households free of problems with rot but who also report poor health status. Another strong result is found, with consistently higher levels of poor health found amongst households with rotten windows. Mean incidences of 12.5% of households (CI=8.2%, 16.9%, α=0.05) and 7.7% of households (CI=5.5%, 9.9%, α=0.05) are found across EU-14 for poor health status and household rot against poor health status and no household rot. T-tests show that this relationship is statistically significant, albeit not quite as strongly as previously (P=0.001), while a correlation coefficient of 0.94 (P<0.001) indicates the consistency of the association between increased incidence of poor health and presence of household rot.

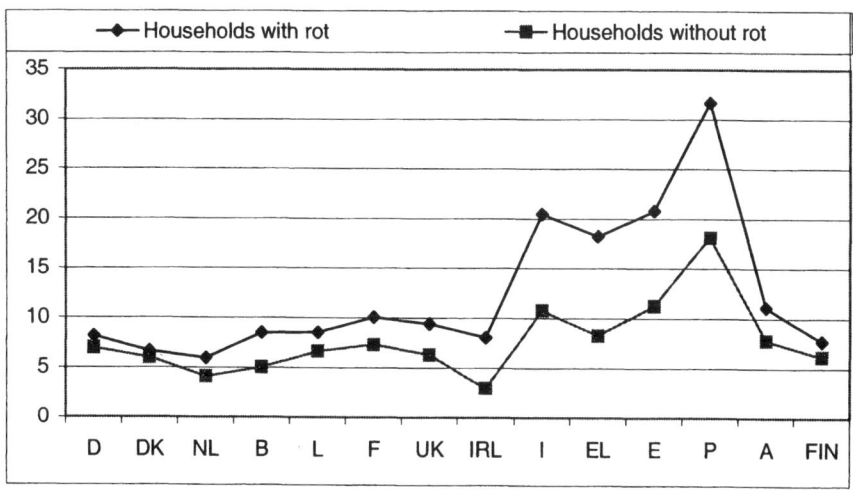

Figure 5.4 Poor health and rot (mean % of households, 1994-97)

The largest percentage-point variations in poor health status are found in Portugal (13.5%) and Greece (10.0%), with the former also reporting the highest overall level of ill health (some 31.7% of households with rotten windows are declaring either very poor or poor health). Ireland, Greece and Italy again demonstrate the largest proportionate increase in the incidence of poor health; those with rotten windows are reporting incidences of poor health status 168%, 120% and 91% (respectively) more than those without rotten windows.

Central Heating

Households not possessing central heating or similar heating systems generally find it more difficult to efficiently heat the home than households that do possess such systems. The lack of either central heating or electric-storage heating is a potentially good objective indicator of fuel poverty and a very good indicator of housing deprivation.

The results of this analysis show that households lacking such heating systems in Europe are persistently reporting higher incidences of poor health than those households equipped with central or electric-storage heating. A mean incidence of poor health of 6.9% (CI=5.2%, 8.5%, α=0.05) of households is calculated for those possessing central and electric-storage heating, compared with 12.3% (CI=9.6%, 15.0, α=0.05) of households lacking such heating equipment. T-tests show that this is a significant result (P<0.001), while a correlation coefficient of 0.95 (P<0.001) indicates that there is a high level of consistency in the association between increased incidences of poor health among households without central-heating systems compared to those possessing such systems.

The largest variations (in percentage points) in poor health status are found in Portugal (8.7% difference in incidence between those with and without central heating), Italy (8.3%) and Greece (7.4%). Once again, the highest incidence of poor health is found in Portuguese households, with 23.7% of households lacking central or electric-storage heating suffering from either poor or very poor health. The largest variations proportionately in poor health status are found in Greece (121% increase in incidence among households with no central heating), Ireland (116%) and Luxembourg (115%).

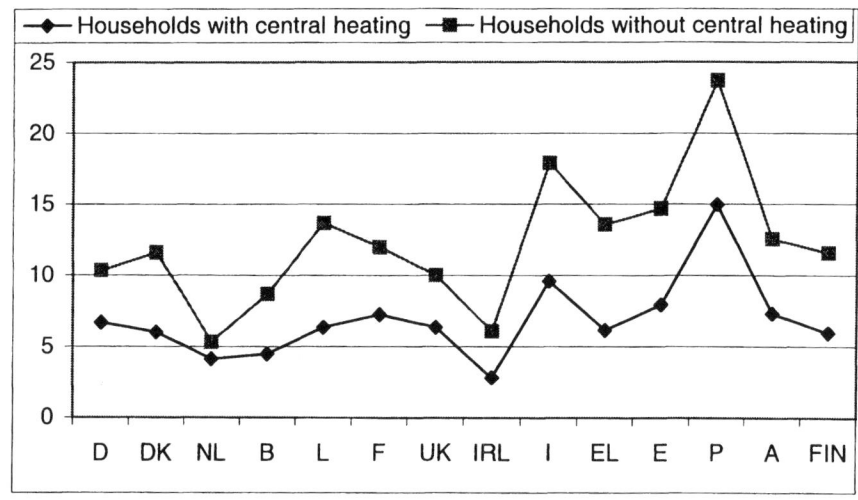

Figure 5.5 Poor health and central heating (mean % of households, 1994-97)

Leaky Roofs

A leaking roof has a number of adverse impacts on households' well-being. Besides the obvious implications for energy efficiency (and the associated excess fuel bills), a leaking roof may cause damp and mould spores to develop in the dwellings' walls. Such spores are especially pernicious to human health, especially for the very young and the elderly, and can lead to respiratory conditions, such as bronchitis and asthma (March *et al.*, 1999). In addition, 82% of British households regard the ability to maintain the home in a decent state as essential, while 93% regard a home free of damp as a necessity. As such, the presence of a leaky roof acts as a solid indicator of housing deprivation, as it appears to be a strong socially perceived necessity.

This study calculates that 12.6% of households with leaky roofs are reporting "poor" or "very poor" health status across Europe (CI=8.1, 17.1, α=0.05), compared with 8.0% of households not demonstrating this adverse housing condition (CI=5.6, 10.3, α=0.05). T-tests show that these results are significant at the 5% level (P<0.001). A very strong correlation coefficient is found for the two sub-samples (R=0.96, P<0.001) with only Germany breaking the trend, showing an insignificant variation in poor health between households with and without leaking roofs. However, it should be noted that just 3.1% of German households suffer from this housing condition in 1997 (Table 2.6).

The largest percentage-point variations are found in Portugal (14.3% more households without adequate heating facilities are suffering from poor health status than those with adequate facilities), and Spain (10.4% variation). The highest level of poor health is found in Portugal, where 34.2% of households without adequate heating facilities self-report poor health. Ireland displays the largest proportionate increase in the incidence of poor health among households with leaking roofs (138%), followed by Luxembourg (95%) and Greece (84%).

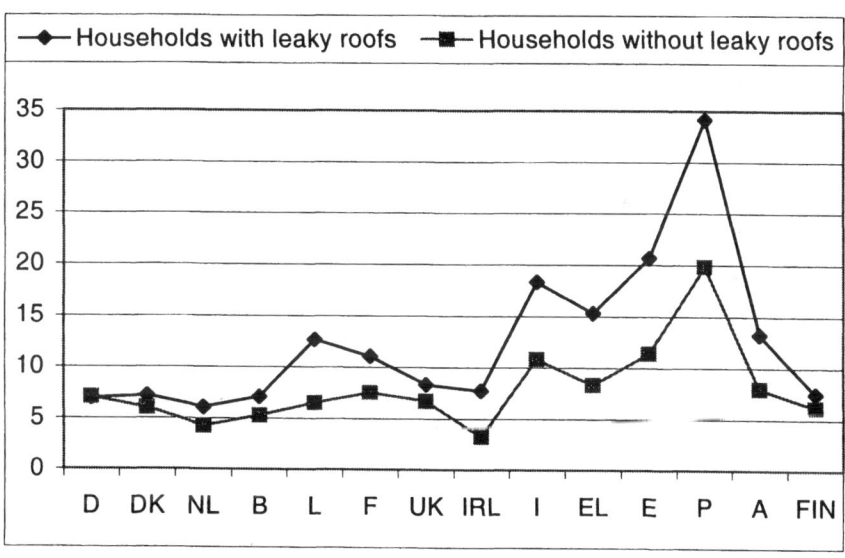

Figure 5.6 Poor health and leaky roofs (mean % of households, 1994-97)

Overcrowding

Household overcrowding is considered undesirable for human health, as it has been shown to be linked with increased rates of various viral and bacterial infections, especially in the respiratory tract (Marsh *et al.*, 1999). There is a long-standing debate about the relative importance of household crowding and social problems, and the literature showing the relationship between crowding and ill health reports somewhat mixed results. Household overcrowding, when used as a socio-economic indicator, has often been defined as households with more than 1 person per room (though this definition may be increased to more than 2 persons per room (Howden-Chapman *et al.*, 1996) for more conservative results); overcrowding has also been calculated in self-reporting data by asking householders whether they have enough space to meet their needs (Healy, 2001b). The levels of overcrowded households may be considered to be useful indicators of housing conditions and, more generally, quality of life across EU-14, however their association with health outcomes remain inconclusive. This analysis tests whether overcrowded households in Europe exhibit significantly higher levels of poor health status than uncrowded households.

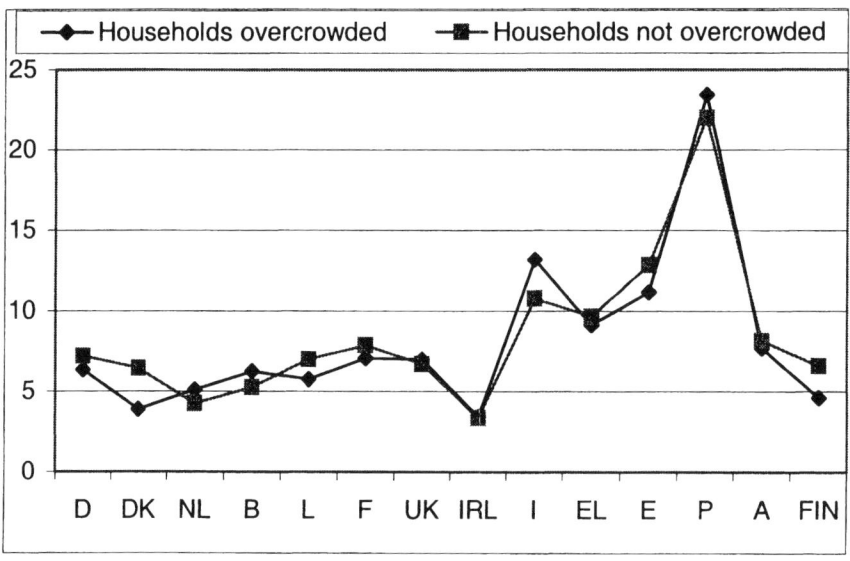

Figure 5.7 Poor health and overcrowding (mean % of households, 1994-97)

Households in this analysis are asked whether they have enough space to meet their needs or are, otherwise, 'overcrowded'. The results are interesting, as no clear relationship is found using the self-reported datasets. In fact, a slightly higher incidence of poor health is found among uncrowded households (8.5% compared with 8.1% across EU-14). Country-to-country variations in poor health are not significant across Europe (P=0.40) and neither percentage-point nor proportionate variations are significant. This is a notable finding, and it goes against recent research in the UK by Marsh et al. (1999) and the findings of Kearns and Smith (1993) for New Zealand households, both of which showed adverse health implications of living in overcrowded environments. It is likely that a number of factors result in unclear findings in this section. *Inter alia*, households' ability to accurately self-report overcrowding is one potential issue, and household composition also plays a factor.

Health and Housing Affordability / Satisfaction

Housing Costs

Households were asked to consider their income and their expenditure on housing, i.e. their mortgage or rent payments, and declare whether or not they found these costs to be financially burdensome. Affordability is generally considered fundamental to housing satisfaction, and there is a considerable literature demonstrating the relationship between housing finances and levels of satisfaction using mainly self-reported survey data.[4] As such, it is considered useful to test if the association between housing affordability and satisfaction may be extended to health outcomes.

Households with heavily burdensome housing costs are found to be associated with increased incidences of poor health status across EU-14, with the exception of Greece, where the pattern is reversed. An overall average incidence of poor health of 12.7% (CI=9.2%, 16.3%, α=0.05) is calculated for households whose housing costs are heavily financially burdensome, compared with just 7.2% (CI=4.8%, 9.5%, α=0.05) of those households whose housing costs are easily met. T-tests reject a null-hypothesis scenario with ease (P<0.001), although a somewhat lower correlation coefficient is found between the two sub-samples (R=0.75, P=0.002), indicating a less-than-perfect fit in the model; this is attributable, to a large degree, to the Greek data.

Large variations in self-reported poor health are found in Luxembourg (13.4%) and Portugal (12.2%). Once again, the highest overall level of poor health is reported for Portuguese households, with 30.3% of households whose housing costs are financially burdensome declaring poor or very poor health. In addition, there is a very large proportionate increase in the incidence of poor health among those with burdensome housing costs in the Netherlands (some 3 times that found among households not declaring burdensome costs), while Luxembourg and the UK also demonstrate large proportionate increases (250% and 130% respectively).

---

[4] A very interesting study by Konadu-Agyemang (2001) identifies lack of affordable housing as a key factor in the "dismal housing situation" facing prospective home-owners in Accra.

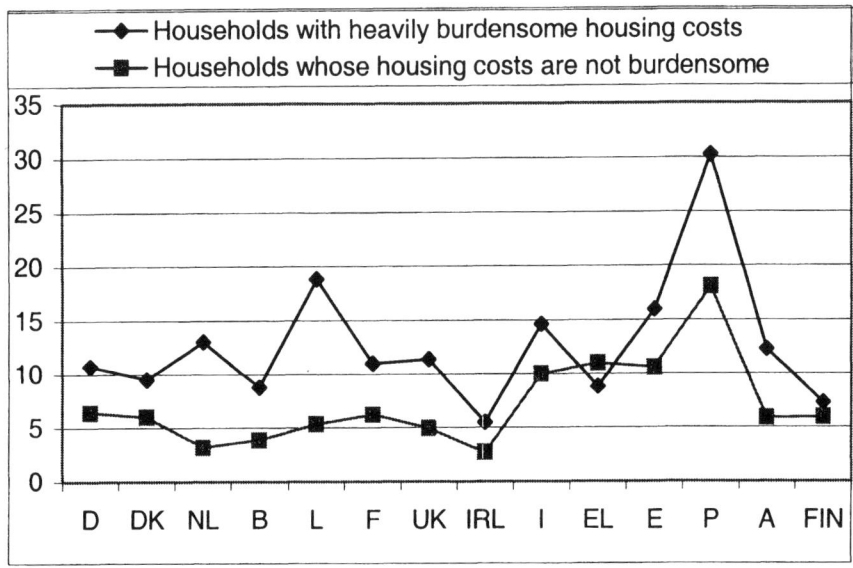

Figure 5.8 Poor health and burdensome housing costs (mean % of households, 1994-97)

Utility Bills

A household which has been unable to pay on time a scheduled utility (gas or electric) bill over the previous 12 months is most likely finding it difficult to keep the home adequately heated and, as such, this affordability indicator also assists in identifying fuel poverty. Those who were unable to keep up to date with their utility bills may also have suffered disconnection from the supplier, compounding the experience of fuel poverty.

The results in this chapter indicate that poor health is associated with an inability on the part of the household to pay on time their scheduled utility bills. About 12.8% (CI=9.1%, 5.5%, α=0.05) of households unable to do this are self-reporting poor health status, compared with 7.6% (CI=5.5%, 10.9%, α=0.05) of households able to afford their utility bills and t-tests show that these variations are highly significant (P<0.001). A high correlation coefficient (R=0.85, P<0.001) demonstrates that there is a high degree of consistency in the self-reported heath data, with all countries reporting higher incidences of poor health among households that experience difficulties in paying their utility bills as scheduled.

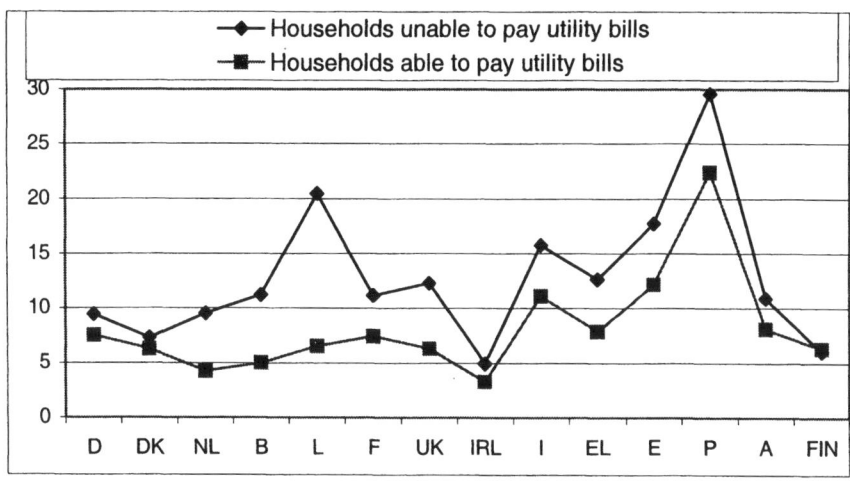

Figure 5.9 Poor health and late payment of utility bills (mean % of households, 1994-97)

Large variations in the incidence of poor health are calculated in percentage-point terms for Luxembourg (13.7%) and Portugal (7.2%). Portuguese households unable to pay utility bills on time demonstrate the highest overall level of poor health, with 29.6% affected. In proportionate terms, Luxembourg appears to have the highest variation in the incidence of poor health among households declaring an inability to pay utility bills, with an incidence of 210% more than households that can meet utility bills as scheduled. Both Belgium and the Netherlands also display similar health outcomes, with an increase in the incidence of poor health of 124% respectively.

Housing Satisfaction

In this section households are asked to rate their satisfaction with their housing on a scale of 1 ("completely satisfied") to 6 ("completely dissatisfied"). The proportion of households who are either "completely dissatisfied" or "very dissatisfied" (5 or 6 ratings) are combined as a measure of housing dissatisfaction, and households either "completely satisfied" (1) or "very satisfied" (2) are combined as a measure of satisfaction with housing. Self-reported health data are compared across the sample. These results are considered important, as housing is a fundamental quality-of-life indicator. The link between housing satisfaction and happiness is well-researched and highly correlated; indeed, it has been shown by Barresi *et al.* (1984), amongst others, to be the key factor influencing happiness, especially in older populations. It was thought beneficial, therefore, to test the link between happiness, housing satisfaction and poor health status, as such a result may have strong policy implications.

The results are remarkable in terms of their magnitude, as they show a very powerful relationship between satisfaction with housing and health status. Across EU-14, repeatedly higher incidences of poor health are reported among dissatisfied households for all countries considered in the analysis. A mean incidence of 14.2% (CI=10.1%, 18.2%, α=0.05) of households "completely" or "very" satisfied with their housing report poor health, compared with 30.5% (CI=22.1%, 38.8%, α=0.05) of households "completely" or "very" dissatisfied with their housing conditions. T-tests show that this variation is highly significant (P<0.001), while a correlation coefficient of 0.91 (P<0.001) demonstrates the consistency of the relationship between the two self-reported health datasets.

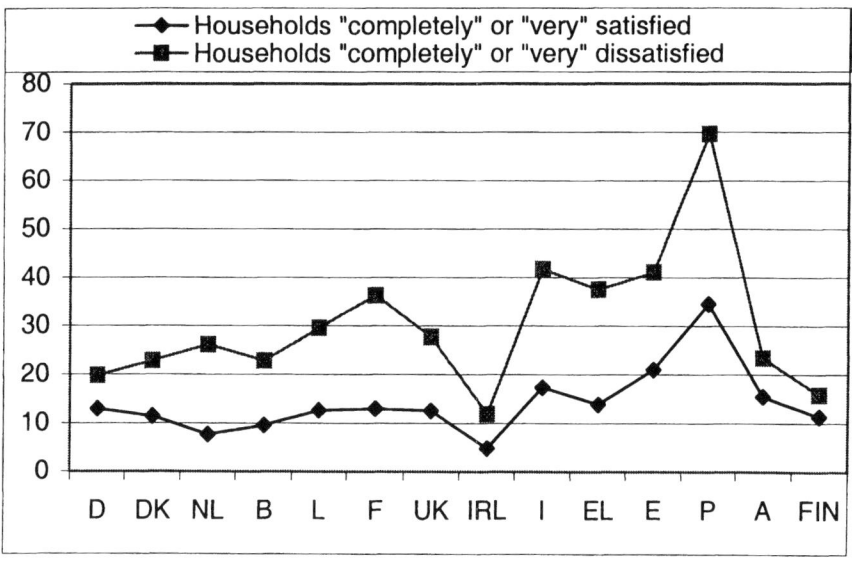

Figure 5.10 Poor health and housing satisfaction (mean % in each group, 1994-97)

The largest variations in poor health are calculated for Portugal, where a difference of 35.2% is found, followed by Italy (24.3%), Greece (23.6%) and France (23.3%). One of the most startling results in this analysis relates to the level of poor health found among Portuguese households "completely" or "very" dissatisfied with their housing; a remarkable 69.7% of these households are demonstrating poor health status which is a serious cause for concern. Such levels of poor health are over twice the average rate found in the dissatisfied households' group across EU-14 and some five times the incidence calculated for satisfied households across Europe. In proportionate terms, variations in health are most notable in the Netherlands, where dissatisfied households demonstrate an increase in the incidence of poor health of 240% above that among satisfied households. France and Greece also display large proportionate variations in health outcomes (incidence of poor health in dissatisfied households 180% and 171% more than satisfied households respectively).

Conclusions

This chapter has presented an analysis of the association between health and a variety of social indicators of housing deprivation in Europe using the first standardised, longitudinal, pan-European dataset on social indicators. This quantitative analysis has shown that there are a number of serious causes for concern. The first major conclusion pertains to the strong relationship found between housing conditions and health. A number of social indicators relating to housing deprivation were employed to assess the health effects of lacking various socially perceived necessities. The results show that, with the exception of overcrowding (for which no relationship is found), all other (16) indicators are found to be significantly associated with increased incidence of poor health status in EU-14. The (in)ability to adequately heat the home – a key socially perceived necessity and indicator of fuel poverty – is found to result in particularly higher incidences of poor health status in the Member States analysed, with over twice the level of poor health reported among fuel-poor households. It was also found that the least energy efficient housing stocks tend to suffer from the highest levels of poor health. Generally, the highest levels of aggregate poor health were found in southern Europe, where housing conditions (regarding both thermal efficiency and general housing deprivation) appear to be the poorest. In northern Europe Austria, France, Luxembourg and the UK demonstrate high incidences of poor health among deprived households and those with below-par housing conditions. Fig 5.10 summarises the statistical results of the chapter.

Housing affordability is also found to be related to health status in EU-14, with significantly increased levels of poor health reported among those with financially burdensome housing costs and those unable to pay utility bills on time. In addition, poor health is found to be highly associated with housing dissatisfaction, with 30.5% of dissatisfied households in EU-14 declaring poor or very poor health; the incidence of poor health of 69.7% of dissatisfied households in Portugal is especially worrying (Figure 5.10).

It is important to bear in mind that, while the incidence of poor health is important across EU-14 in identifying aggregate health outcomes, it is perhaps more useful to compare the variations in the incidence of poor health associated with various social indicators, as this indicates more precisely the health effects of a given deprivation indicator. This is because poorer countries (like Greece, Spain and Portugal) are likely to report higher incidences of poor health regardless of whether or not they declare deprivation indicators. Some distinct patterns emerge when proportionate variations in health are examined across EU-14. Ireland appears to suffer from the largest proportionate increases in poor health among households demonstrating an indicator of housing deprivation; Greece and Italy also suffer from large proportionate increases in poor health with regard to housing deprivation and fuel poverty. The Netherlands suffers persistently from the largest proportionate increases in poor health with regard to housing affordability and satisfaction with housing, while Luxembourg also performs poorly with regard to the former.

Table 5.1 Summary of statistical analysis on deprivation indicators and poor health

Indicator	'Yes' mean %	'Yes' 95% CI mean %	'No' mean %	'No' 95% CI mean %	P-value
Heat	15.5	12.9, 18.2	6.8	5.5, 8.1	<0.001
Facil	12.0	9.1, 14.9	7.4	5.4, 9.5	<0.001
Damp	12.1	8.0, 16.2	7.6	5.5, 9.8	<0.001
Rot	12.5	8.2, 16.9	7.7	5.5, 9.9	0.001
C/H	12.3	9.6, 15.0	6.9	5.2, 8.5	<0.001
Leak	12.6	8.1, 17.1	8.0	5.6, 10.3	<0.001
O/C	8.1	5.2, 11.1	8.5	5.8, 11.1	0.40
Bills	12.8	9.1, 5.5	7.6	5.5, 10.9	<0.001
Costs	12.7	9.2, 16.3	7.2	4.8, 9.5	<0.001
Dissat	30.5	22.1, 38.8	14.2	10.1, 18.2	<0.001

Note: 'Heat'=Unable to heat home adequately, 'Facil'=Inadequate heating facilities, 'Damp'=Presence of damp in the home, 'Rot'=Presence of rot in the home, 'C/H'=No central heating, 'Leak'=Leaky roofs, 'O/C'=Overcrowding, 'Bills'=Unable to pay utility bills on time, 'Costs'=Housing costs are heavily financially burdensome, 'Dissat'=Very or totally dissatisfied with housing conditions

Table 5.2 Correlation coefficients and significance levels for deprivation levels and poor health

Indicator	R	P-value
Heat	0.75	0.002
Facil	0.92	<0.001
Damp	0.95	<0.001
Rot	0.94	<0.001
C/H	0.95	<0.001
Leak	0.96	<0.001
O/C	0.97	<0.001
Bills	0.85	<0.001
Costs	0.75	0.002
Dissat	0.91	<0.001

Although many results appear to corroborate the majority of research regarding the relationship between housing deprivation and health, there are some apparent anomalies. Despite having among the least energy efficient housing stocks in northern Europe (Table 2.12) and among the highest levels of income poverty, and inequality in Europe (NESF, 2000), Irish households persistently report among the lowest absolute levels of poor health in EU-14. A number of interpretations are possible, however one plausible explanation is that the Irish, either intentionally or otherwise, 'under-declare' their self-perceived levels of poor health. Cultural idiosyncrasies always play a part in self-reported results, and it is possible that they affect the (relatively low) magnitude of absolute levels of poor health in Ireland. The findings regarding the lack of a relationship between health and overcrowding also require some additional comments. There is a considerable medical and social-policy literature highlighting the effects on health of crowding in the home. However, much of this research uses objective measures of overcrowding (as opposed to subjective, self-reporting measures). It is possible that the results (especially those based on subjective indicators) are showing some interplay across EU-14 owing to cultural factors. Many households may feel embarrassed to declare shortage of space, while others may not regard their housing conditions as crowded when, using an objective measurement, such households may be defined as objectively overcrowded.[5]

---

[5] Healy (2002a), in a cross-country analysis of housing conditions in Europe, showed that Irish (and Italian) households continuously under-declare their perceived levels of household crowding, despite objective measures of overcrowding reporting otherwise.

Chapter 6

Fuel Poverty and Health in Ireland

Introduction

The impaired health status of fuel-poor households in Europe was demonstrated in a previous chapter. Data from the Urban Institute of Ireland's (UII) national household survey of Ireland are used in this chapter to analyse, in more detail, specific health outcomes associated with fuel poverty in Ireland. In light of the key policy-drivers associated with improved energy efficiency, outlined in Chapter 5, this study identifies some key associations between fuel poverty and health using self-reported data. The context for Ireland has already been discussed in previous chapters, so this chapter concentrates on presenting the results of the UII survey; some discussion is given towards the end of the chapter regarding the key findings of the survey.

Results: Fuel Poverty and Health

This section presents the results of a national household survey conducted in 2001 to elicit a variety of information on social indicators and living conditions in Ireland. The survey was funded by the Urban Institute of Ireland, University College Dublin. It uses a random (probability-based) sample of 1,500 households across Ireland to yield the required data; such a sample is statistically significant in a population of 3.7 million people and 1.3 million dwellings. The definition of fuel poverty employed in this study is the basic qualitative definition first used by Lewis (1982) and later modified by Clinch and Healy (1999a): 'the inability to heat the home adequately because of low household income and energy inefficient housing'.

Households reporting an inability to heat the home to an adequate, comfortable temperature are, thus, deemed fuel-poor by definition. In this chapter the relationship between fuel poverty and the following health outcomes is analysed:

- Self-reported health status (using a five-point response variable)
- Self-reported health status now compared to a year ago (using a five-point response variable)
- Households demonstrating various manifestations of poor physical health
- Households demonstrating various manifestations of poor mental or emotional health

- Feelings
- Self-perceived health status
- Objective health outcomes (e.g. number of visits to their GP, A&E admittances, etc.)
- Chronic health outcomes (such as asthma, hypertension and arthritis)
- Housing and chronic health outcomes: identifying self-perceived causality
- Worries (regarding both personal physical and emotional health and that of their children and other related variables)
- Quality of life and life satisfaction as a whole

Results are only reported in this study where sample sizes allow for a reasonably low level of the margin of error. Where frequency counts are too small to allow for reliable comparisons, a dash, denoted '-', appears in the data tables. The chapter does not devote space to detailed statistical or econometric analyses of the results.[1] Rather, for the sake of clarity and brevity, the chapter presents key cross-tabulation statistics relating to the associations (or lack of them) found between fuel-poor households and various related health outcomes. Households in the survey are first split into two discrete groups: fuel-poor households (those declaring an inability to adequately heat the home to a comfortable temperature) and other (non-fuel-poor) households. Some 17.4% of the sample (representing 226,000 households) report some level of fuel poverty in Ireland.

Self-Reported General Health Status

Table 6.1 reports the first set of results in this chapter. It is clear that there is a strong relationship between self-reported health status and fuel poverty. Some 20.5% of households with no problem of fuel poverty in Ireland report excellent health, compared with just 10.8% of those living in fuel poverty. A significant result is also found for those reporting "very good" health status (41.2% compared with 34.8%). Conversely, the incidence of self-reported poor health status is higher among fuel-poor households, though not to quite the same magnitude as the variations among those with good health, with 4.3% affected compared 2.0%. Fuel-poor households are much more likely to report "fair" health status than those not suffering from fuel poverty, with 18.2% of all fuel-poor households declaring such health compared with just 7.4% of those households unaffected. Nonetheless, about three-quarters of fuel-poor households report "good" or better health, which reduces the magnitude of the above findings.

---

[1] A detailed, multivariate, Probit regression analysis of fuel poverty can be found in chapters 5 and 6.

Table 6.1 Fuel poverty and self-reported health status

Currently...	Fuel-poor % households	Not fuel-poor % households
Excellent	10.8	20.5
Very good	34.8	41.2
Good	32.0	28.7
Fair	18.2	7.4
Poor	4.3	2.0

Self-reported Health Status Now Compared to Last Year

It is clear that fuel-poor households believe their health status is, in general, worse now than compared to 12 months ago. Some 10.9% of fuel-poor households believe this to be the case, stating that their health is either "somewhat worse now" or "much worse now" compared to a year ago, compared with 5.2% of non-fuel-poor households. Thus, the data appear to be demonstrating something of a dynamic effect of fuel poverty on health status. However, it should be noted that 12.8% of fuel-poor households actually report improvements in health now compared to one year ago. Exogenous factors pertaining to the healthy state of the Irish economy and improved living standards are, undoubtedly, playing a part in these positive trends.

Table 6.2 Fuel poverty and self-reported health status now compared to 1 year ago

Compared to 1 year ago...	Fuel-poor % households	Not fuel-poor % households
Much better now	2.3	2.7
Somewhat better now	10.5	10.7
About the same	76.3	81.3
Somewhat worse now	8.2	4.5
Much worse now	2.7	0.7

Manifestations of Poor Physical Health

In this section of the survey, households are asked about whether their physical health has impacted on their ability to work or carry out other activities over the past month. Significant variations are found between the fuel-poor and those unafflicted. For each of the four categories of manifestations of poor physical health examined, the proportion of fuel-poor households that declares health-related problems is at least twice that found among non-sufferers. The first question on this section of the UII national survey asks households if they have had to cut time spent working or undertaking other activities in the past month because of poor physical health. Some 13.9% of fuel-poor households responded affirmatively, compared to just 5.9% of non-suffering households. Similarly, the proportion of households that accomplished less than they would have liked owing to poor physical health is substantially higher among households declaring fuel poverty (14.7% against 6.0%). Almost identical percentages of households report limitations in the kind of work or activities undertaken over the past four weeks because of ill health (14.7% and 5.9% respectively). Finally, 15.8% of fuel-poor households declare that their work or other activities undertaken over the past month took extra effort because of their poor health status, compared with 6.7% of households unaffected by fuel poverty.

Table 6.3 Fuel poverty and manifestations of poor physical health

Over the past 4 weeks…	Fuel-poor % households	Not fuel-poor % households
Cut time spent working/other activities	13.9	5.9
Accomplished less than would have liked to	14.7	6.0
Limited in the kind of work/activities done	14.7	5.9
Work/activities took extra effort	15.8	6.7

Manifestations of Poor Mental Health

Very similar results are found with regard to manifestations of poor mental/emotional health. The same four outcomes are analysed, and typical proportionate variations between sufferers and non-sufferers of fuel poverty are between 100% and 150%. Some 13.9% of fuel-poor households have cut time spent working or carrying out other activities over the past four weeks owing to their poor emotional health, compared with 5.9% of households able to heat their home adequately. It is found that 15.1% of fuel-poor households accomplished less than they would have liked over the past month because of poor emotional or mental health, while the same proportion were limited in the type of work or other activities done; non-sufferers of fuel poverty report incidences of 6.0% and 5.9% respectively. Finally, 15.8% of the fuel-poor stated that their work or other activities took extra effort due to impaired emotional or mental health over the past month, compared with 6.7% of those able to heat their homes adequately. While the data tell little about the effect of fuel poverty on health, they do demonstrate the increased risk of being physically limited which is associated with being fuel-poor.

Table 6.4 Fuel poverty and manifestations of poor emotional health

Over the past 4 weeks…	Fuel-poor % households	Not fuel-poor % households
Cut time spent working/other activities	13.9	5.9
Accomplished less than would have liked to	15.1	6.0
Limited in the kind of work/activities done	15.1	5.9
Work/activities took extra effort	15.8	6.7

Feelings

The national household survey elicited information regarding personal feelings of individual householders. These relate to whether they feel 'full of life', calm, happy, depressed, lethargic, worn out, and so forth. A six-point response variable recorded the magnitude of such feelings; discrete responses ranged from 'all of the time' to 'none of the time' with graduated variations between both extremes. When the results are cross-tabulated with households declaring fuel poverty, some interesting associations are found, with proportionate variations of up to 300% between the fuel-poor and non-fuel poor. The strongest variation occurs with householders 'never' or 'rarely' full of life, with a quarter (24.7%) of fuel-poor households stating this feeling compared with just 8.3% of non-sufferers. Some 24.4% of fuel-poor households declare severe lethargy ('never' or 'rarely' having lots of energy) compared with 10.2% of non-sufferers. Exactly one-in-ten fuel-poor households state that they 'never' or 'rarely' feel calm and peaceful, compared with 5.1% of households unaffected by fuel poverty. Similar variations are found for households 'always' or 'mostly' tired (9.7% versus 4.7%). Less marked variations are found for outcomes relating to general happiness, feeling low or downhearted, feeling warn out, feeling nervous and feeling depressed.

Table 6.5 Fuel poverty and feelings

Over the past month…	Fuel-poor % households	Not fuel-poor % households
Never/rarely feel 'full of life'	24.7	8.3
Never/rarely feel calm and peaceful	10.0	5.1
Never/rarely have lots of energy	24.4	10.2
Never/rarely happy	7.8	5.3
Always/mostly downhearted and low	3.9	2.2
Always/mostly worn out	7.0	4.2
Always/mostly very nervous	3.1	2.3
Always/mostly so depressed that nothing could cheer me up	3.9	2.3
Always/mostly tired	9.7	4.7
Health has always/mostly limited my social activities	5.4	2.5

Self-Perception of Health

It is useful to examine whether households believe they are relatively healthier or less healthy than average. This is particularly useful in identifying whether fuel-poor households are aware of their increased risk of impaired health status. The results show a clear relationship between fuel poverty and self-perceived health status. Exactly 11% of fuel-poor households believe it is 'definitely' or 'mostly' correct to say that they fall ill more easily than other households, while just 4.5% of households able to afford to heat their homes adequately believe this to be the case. Some 15.6% of fuel-poor households expect their health to deteriorate in the near future, compared with just 6.1% of non-fuel-poor households. Exactly 16% of fuel-poverty sufferers believe that the statement "I am as healthy as anyone else I know" is 'definitely' or 'mostly' false, compared with 8.2% of non-sufferers. Finally, 18.4% of those living in fuel poverty believe that it is 'definitely' or 'mostly' false to state that their health is excellent, compared with 10% of households not declaring fuel-poverty problems.

Table 6.6 Fuel poverty and self-perceived health status

In general...	Fuel-poor % households	Not fuel-poor % households
"I seem to get ill more easily than others": 'Definitely' or 'mostly true'	11.0	4.5
"I expect my health to get worse": 'Definitely' or 'mostly true'	15.6	6.1
"I am as healthy as anyone I know": 'Definitely' or 'mostly false'	16.0	8.2
"My health is excellent": 'Definitely' or 'mostly false'	18.4	10.0

Objective Health Outcomes

A number of objective health outcomes are now analysed. These have the advantage of being less open to biases associated with self-reporting of subjective data regarding feelings and self-perceived health, etc. The objective health outcomes in this chapter relate to whether the household:

- was admitted to hospital as a day-case in the past year
- was admitted to hospital as an outpatient in the past 3 months
- attended an Accident and Emergency clinic in the past 3 months
- sought the advice of a chemist in the past three months
- suffers from a long-standing illness, disorder or disease

In addition, a high number of visits to a GP in the past year (six or more) is also employed as an adverse health indicator. All of these variables are cross-tabulated with households' ability to heat the home adequately. The results, reported in Table 6.7, show some interesting associations with fuel poverty.

Perhaps the most worrying finding relates to households declaring long-standing illnesses and diseases. One-in-five (19.8%) fuel-poor households report such a health problem, compared with just 10.5% of fuel-rich households; in other words, a proportionate variation of 89% is found between the fuel-poor and other households. This is a disturbing finding, as it indicates that a significant proportion of individuals suffering from a variety of serious, chronic conditions (ranging from asthma to ischaemic heart disease) may be living in sub-standard housing conditions where the inadequate level of warmth may exacerbate their symptoms.[2] A number of epidemiological studies have, in the past, demonstrated a strong relationship between cold, damp conditions and increased respiratory and cardiovascular illnesses, and this study corroborates this relationship using macro data.[3] While this study cannot conclusively identify causality (i.e. does fuel poverty directly result in more of these diseases or are fuel-poor households more likely to contract such diseases for other reasons?), the results are, nonetheless, policy relevant and of direct concern to public health and housing policy. Because of the potentially serious nature of this result, the next section explores this finding in far more detail, isolating the chronic diseases associated with fuel poverty and inadequately heated homes.

In addition to the above finding, some 17.8% of households reporting fuel poverty have visited their GP on six or more occasions in the past year, compared with 8.0% of households unaffected by fuel poverty – a 123% proportionate variation. Similarly strong variations are found for households who have attended outpatients departments in the past three months (21.3% compared with 10.8%), Accident and Emergency (or casualty) departments (11.2% versus 7.2%) and hospital admission as a day-case (13.3% against 7.2%). Furthermore, two-fifths (40.1%) of fuel-poor households have sought the advice of their chemist in the past three months, compared with 26.5% of other households.

Table 6.7 Fuel poverty and objective health outcomes

Health outcome measure	Fuel-poor % households	Not fuel-poor % households
≥ 6 visits to GP in past year	17.8	8.0
Outpatient in past 3 months	21.3	10.8
A&E attendance in past 3 months	11.2	7.2
Hospital admission in past year	13.3	7.2
Chemist's advice in past 3 months	40.1	26.5
Long-standing illness/disorder	19.8	10.5

---

[2] Although it should be noted that this is an inference, as no direct causality has been shown in this chapter.
[3] See Marsh *et al.* (1999) and Rudge and Nicol (2000) for summaries of such research.

Chronic Health Outcomes

In order to address the finding that fuel-poor households are twice as likely to suffer from a long-standing illness or disease than non-suffering households, this section identifies the specific diseases associated with fuel poverty. The following chronic conditions, all of which have been either shown or hypothesised to be associated directly with chronic cold exposure, were specified in the national survey:

- Ischaemic heart disease
- Cerebrovascular disease
- Pneumonia
- Asthma
- Emphysema
- Hypertension
- Other cardiovascular problems
- Other respiratory problems
- Arthritis
- Depression
- Headache

Unfortunately, because of the low frequency of households in each of the disease categories, only certain results are statistically significant. Thus, only those that are based on a reasonable sample size are reported. The survey finds that fuel-poor households are up to four times as likely to suffer from specific chronic conditions than non-fuel-poor households. The most dramatic variations occur with respiratory conditions other than asthma, where a remarkable 27.5% of households living in fuel poverty are reporting such conditions, compared with 7.1% of other households. Arthritic conditions have traditionally been relatively prevalent in Ireland, and this survey demonstrates increased incidences among the fuel-poor with exactly two-thirds (66.7%) of such households suffering from arthritis. Although 35.3% of fuel-poor households report problems with high blood pressure, a similarly high proportion (33.1%) of those able to adequately heat their homes also report hypertension. Equally, no significant variation is found for ischaemic heart disease (though the sample size in the survey was modest) or headaches. However, a strong association with fuel poverty was found for chronic depression, where one-in-five households reported such a condition, while just 7.1% of non-sufferers reported same.

Again, these results do not prove that fuel poverty is a causal factor in increased health; rather, they indicate that the fuel-poor have higher incidences of chronic health problems. The data are also far from conclusive in identifying this latter effect, as was noted earlier. In fact, the data only reveal significant variations between fuel-poor households and other households for chronic disorders such as arthritis, depression and respiratory problems other than asthma and emphysema.

Table 6.8 Fuel poverty and related chronic health outcomes

Chronic health outcome	Fuel-poor % households	Not fuel-poor % households
Ischaemic heart disease (heart attack)	7.8	5.5
Cerebrovascular disease (stroke)	-	-
Pneumonia	-	-
Asthma	15.7	13.4
Emphysema	-	-
Hypertension (high blood pressure)	35.3	33.1
Other cardiovascular problems	15.7	12.6
Other respiratory problems	27.5	7.1
Arthritis (and related joint problems)	66.7	45.7
Depression	19.6	7.1
Headache	17.7	16.5
Self-perceived ill health caused by poor housing	25.0	9.2

Health and Housing: Self-Perception

This section of the chapter examines a crucial question. Respondent households are asked to consider whether they believe their chronic health problems are affected by their housing conditions. A quarter (25.0%) of fuel-poor households believe that their housing conditions played a causal role in their chronic illness, while just 9.2% of other (non-fuel-poor) households believed this to be the case (Table 6.8). It is of course likely that a considerable number of households do not realise the linkage between thermally inefficient, cold housing and increased levels of poor health. As a result of this 'information gap', it is likely that these results underplay the real self-perceived causal link between housing and health. Nonetheless, even allowing for this important caveat, the result shows that fuel-poor households perceive their housing conditions to be pernicious to their health to a far greater extent than other households, with a 172% proportionate variation in the results between the two household types. It would be very interesting to see how this proportion varies as the information gap closes, which it is likely to do over time as improved energy efficiency and the alleviation of fuel poverty become higher on the political agenda of the Irish government.

Worries

Households were also asked in the national survey about how often they worry about various aspects of their lives including their own health, the health of their children (if applicable), their financial situation, job security and housing. A four-point response vector was utilised which contained the discrete categories 'never', 'a little', 'sometimes' and 'all of the time'. Table 6.9 illustrates the results. It is notable that households, in general, worry to a far greater extent about their physical, rather than their emotional, health; and fuel-poor households in particular (31.0% worry 'all of the time' or 'sometimes' compared with 23.3% of other non-suffering households). Half of fuel-poor households (49.7%) worry about their children's health all of the time or sometimes, while two-fifths (40.5%) of the non-fuel-poor report similar levels of worry about their children's health. A remarkable 58.4% of fuel-poor households worry all of the time or sometimes about their financial situation, compared with 38.6% of non-sufferers. Similar variations are reported for job security (48.4% against 31.0%). The most surprising result relates to housing. It would seem plausible that households living in fuel poverty would tend to worry about their housing situation more so than those able to heat their homes comfortably. This hypothesis is not borne out in the results, which show insignificant variations between the fuel-poor and the non-fuel-poor (25.8% versus 23.9%). Such a result may be due to another information gap relating to households' lack of knowledge of the benefits of an energy efficient house in terms of reduced fuel bills, affordable warmth, increased comfort and improved health status; but it is nonetheless a surprising anomaly. It is possible that fuel poverty is part of, but does not add to, the burden of the multiply deprived.

Table 6.9 Fuel poverty and worries

Households that worry 'all the time' or 'sometimes' about their...	Fuel-poor % households	Not fuel-poor % households
Physical health	31.0	23.3
Mental/emotional health	15.1	13.9
Children's health	49.7	40.5
Finances	58.4	38.6
Housing	25.8	23.9
Job security	48.4	31.0

Quality of life

It has been shown that housing is a strong indicator of quality of life (Barresi *et al.*, 1984; Kozma and Stones, 1983). It would, therefore, be logical to assume that fuel-poor households are more likely to express higher levels of dissatisfaction (and lower levels of satisfaction) with their quality of life and life as a whole. This hypothesis is very much verified using the data from the national household survey in Ireland. Fuel-poor households are less likely to report either 'good' or 'excellent' quality of life than other households. Just 48% of fuel-poor households declare that their quality of life is good, compared with 63.5% of other households. Conversely, over two-fifths of fuel-poor households classify their quality of life as 'fair' or 'poor' compared with 21.1% of other (non-suffering) households.

When asked to consider their life as a whole – a very similar question, though worded differently and with a different six-point (as opposed to four-point) response variable – households respond similarly. Just 30.5% of fuel-poor households believe their life is 'very good' or 'as good as can be' compared with

55.6% of other households. On the other hand, 32.9% of fuel-poor households believe their life is 'alright' or 'bad', compared with just 11.4% of non-sufferers. Thus, up to one-third of fuel-poor households are expressing some level of unhappiness with their life, a worrying result for policy-makers.

Table 6.10 Fuel poverty and quality of life/life satisfaction

	Fuel-poor % households	Not fuel-poor % households
Excellent	10.9	15.3
Good	48.0	63.5
Fair	35.9	19.4
Poor	5.1	1.7
As good as can be	7.8	15.6
Very good	22.7	40.0
Good	34.1	32.3
Alright	28.2	10.2
Bad	4.7	1.2
As bad as can be	1.6	0

Discussion

Some findings of this study are now discussed and some clarifications and caveats are emphasised. The study, as a whole, is inconclusive in demonstrating a link between poor housing and adverse health impacts. Many objective indicators of ill health presented in Tables 6.7 and 6.8 report unclear findings and some subjective results are quite surprising. However, both subjective and objective indicators of health are considered equally important; the former are particularly useful in eliciting experiential health outcomes. As with all self-reported data, it is important to stress that cultural idiosyncrasies often play a part in survey results. It was shown by Healy (2002a) that Ireland and Italy, for instance, continuously under-declared their perceived level of poor health, even when it was shown that generalised health outcomes would have been poorer using objective (non-self-reported) measures. It is likely in this study that much of the results are actually conservative, i.e. some households, for a variety of reasons, may not wish to state, for example, that they are suffering from poor emotional or mental health. It is important to take this on board when analysing the data. In this regard, some apparently moderate findings may take on more weighty significance. However, it is clear that this chapter has demonstrated the presence of some levels of reduced health associated with multiple deprivation, though not across all indicators or health outcomes.

Nonetheless, some key conclusions can be made. Fuel-poor households persistently report lower levels of health status (and higher levels of poor or impaired health), however no causal connection can be identified precisely. It was shown, however, that the incidence of excellent health status among fuel-poor households is half that found in all other households. While there appears to be a dynamic aspect to the results (it was demonstrated that the proportion of fuel-poor households believing their health status is impaired now compared to a year ago is over twice the incidence reported amongst all other households), this relationship is far from conclusive. In fact, 12.8% of fuel-poor households report an improvement in their health status over time. There is a discursive medical and social-policy literature dealing with the chronic health aspect of fuel poverty (and chronic exposure to cold, damp housing). These data, on balance, do not attest this dynamic effect.

While the age profile of the fuel-poor is likely to play a part in these results, the data do appear to demonstrate increased hardship among the fuel-poor, though some activities were found to be far more significant than others. All four manifestations of poor health were found to be strongly associated with the fuel-poverty status of the household, and such manifestations were found to result in similar results for both physical and mental/emotional health. With regard to personal feelings, the ability to feel full of life, calm and peaceful, energetic and not lethargic were all found to be negatively associated with the incidence of fuel poverty.

It was interesting to note that many fuel-poor households are self-aware of their health risks and reduced health status. Households living in fuel poverty believe they become ill more easily than on average, they expect their health to disimprove over time, they do not believe they are generally as healthy as the typical person, and they are far less likely to state that their health is excellent. Using objective measures of health status, it was shown that the fuel-poor are more likely to visit their GP regularly. They also demonstrated relationships between other objective health outcomes (e.g. they are more likely to be admitted to hospital as a day-case or to A&E or as an outpatient). Interestingly, fuel-poor households are far more likely to seek the advice of a pharmacist than other households. The key adverse outcome found here relates to the result that fuel-poor households are almost twice as likely to suffer from a chronic illness or disease than other households, with one-in-five affected. The variation is most acute for chronic diseases of the respiratory system and long-term depression.

A major finding of the study relates to housing as a self-perceived causal factor in the levels of poor health status, or more particularly in the levels of chronic diseases. Fuel-poor households are relatively far more likely (three times, in fact) to blame housing conditions as a key cause for their illness than other households. This does not take account of the fact that many people are unaware of the link between housing and ill health and, as such, this result is likely to be a conservative one.

Fuel-poor households worry more than other households. In particular, they worry more about their finances and job security, their children's health, and their own physical health. Curiously, they do *not*, however, worry to a significantly higher degree about their housing situation than other households. As might be expected, quality of life and general life satisfaction is found to be negatively affected by the presence of fuel poverty. Fuel-poor households are more likely to report both poor quality of life and low levels of overall life satisfaction, with proportionate variations of up to 200% between households suffering from fuel poverty and those not enduring such problems.

Overall, there is much inter-correlation between the variables analysed in this chapter, and it is difficult to assert that fuel poverty is *itself* a causative factor in impaired health status. Fuel-poor households often report significantly higher incidences of various adverse health outcomes, but unfortunately this study is unable to identify precise causality. However, the overall thrust of the findings of this chapter indicate that it would be very foolish indeed to state that fuel poverty does not result in an increased risk of adverse health outcomes.

Chapter 7

Fuel-Poor Households and Risk Factors

Introduction

There has been a slowly-growing body of literature developing in the field of fuel poverty and related health effects, especially in the UK (Marsh et al., 1999; Rudge and Nicol, 2000). Despite this, there has been virtually no work examining whether fuel-poor households tend to follow the pattern of those households enduring generalised income poverty and deprivation who exhibit increased levels of a range of risk factors associated with low household income. This research deficit has occurred, not because of a lack of interest in the area, but because of some major logistical reasons, most obviously the lack of data. This chapter attempts to shed some light on this area by analysing the relationship between fuel-poor households and a range of risk factors other than those pertaining specifically to fuel poverty. The data are employed from a statistically representative national household survey of Ireland conducted in 2001 and funded by the Urban Institute of Ireland (UII).

Methods

This section presents the results of the UII national household survey conducted in 2001 to elicit a variety of information on social indicators and living conditions in Ireland. The survey was funded by the Urban Institute of Ireland, University College Dublin and uses a random (probability-based) sample of 1,500 households across Ireland to yield the required data; such a sample is statistically significant in a population of 3.7 million people and 1.3 million dwellings. The definition of fuel poverty employed in this study is the standard qualitative definition first used by Lewis (1982) and later modified by Clinch and Healy (1999a): 'the inability to heat the home adequately because of low household income and energy inefficient housing'.

Households reporting an inability to heat the home to an adequate, comfortable temperature are, thus, deemed fuel-poor by definition. In this chapter fuel-poor households and other (non-suffering) households are analysed to examine whether the former group exhibits increased levels of risk factors other than fuel poverty and attendant indoor cold and damp exposure. There is a growing literature examining fuel poverty and related health impacts, much of which demonstrates a clear association between fuel poverty and increased levels of impaired health status (or lower incidences of good health). However, virtually no research has examined whether such households expose themselves to other exogenous risk factors, such as lifestyle hazards. This chapter attempts to do this by testing for associations with the following:

- Winter time spent outdoors
- Outdoor cold exposure
- Duration of outdoor cold exposure
- Duration of indoor cold exposure
- Lifestyle risk factors
- Diet
- Levels of physical activity
- Usual mode of transport (when shopping)

Results are only reported in this study where sample sizes allow for a reasonably low level of the margin of error. Where frequency counts are too small to allow for reliable comparisons, a dash, denoted '-', appears in the data tables. The chapter does not devote space to detailed statistical or econometric analyses of the results.[1] Rather, for the sake of clarity and brevity, the chapter presents key cross-tabulation statistics relating to the associations (or lack of them) found between fuel-poor households and a range of lifestyle (and other) risk factors associated with ill health. Households in the survey are first split into two discrete groups: fuel-poor households (those declaring an inability to adequately heat the home to a comfortable temperature) and other (non-fuel-poor) households. Some 17.4 per cent of the sample (representing 226,000 households) report some level of fuel poverty in Ireland; such a result is similar to that found recently in the UK of 16.4 per cent in which an expenditure approach to calculating fuel poverty was employed (DEFRA and DTI, 2001).

Results

Duration of Winter Time Spent Outdoors

If fuel-poor households are known to suffer from increased levels of poor health status because of chronic cold (and damp) exposure from within the home, then it is useful to examine whether such households spend more time outdoors during cold spells in the winter than other households. The hypothesis in this section states that the fuel-poor are more likely to spend time outdoors in winter, exposing themselves to potential cold strain, than other households.

---

[1] A multivariate, Probit regression analysis of fuel poverty can be found in the Appendices.

Table 7.1 Fuel poverty and winter time spent outdoors

	Fuel poor % households	Other households %
Virtually no time	11.2	8.1
<10 mins.	6.2	6.7
11-20 mins.	5.0	7.8
21-30 mins.	11.6	14.2
31-40 mins.	8.1	8.7
41-50 mins.	3.9	4.2
51-60 mins.	10.0	9.4
1-2 hours	10.8	16.4
c. 2 hours	17.4	13.9
Half a day	7.7	5.0
Whole working day	6.6	5.6

Table 7.1 reports the results of the national survey in this regard. There appears to be something of an association found between duration of time spent outdoors in the wintertime and the fuel-poor status of the household. In general, fuel-poor households are more likely to spend longer intervals out of doors in the winter than other households, but the association is not strong and mixed results are found for shorter time periods spent outdoors. In any event, such data are of limited use in terms of risk-factor analysis, as there are many confounding variables that might explain such variations.

Outdoor Cold Exposure Resulting in Shivering

Rather than simply identifying duration of time spent out of doors, it is more useful to identify whether certain households protect themselves more adequately from the cold outdoors than others. In this regard, a question was asked on the national survey which attempted to elicit whether fuel-poor households are more or less likely to protect themselves from the cold to avoid shivering episodes. As an outcome gauge, shivering is a useful (and simple) physiological measure which indicates that the body is not adequately protected from the cold environment and is experiencing some level of thermoregulation. While short spells of shivering are relatively harmless among the young, such physiological reflexes are thought to be more pernicious to older individuals, often leading to cardiovascular strain and potentially other, more severe manifestations in the form of adverse conditions such as hypothermia and pneumonia (Collins, 1986).

Table 7.2 presents the results of this question. Fuel-poor households are found to be substantially more likely to feel cold and shiver when outdoors on a cold winter day than other households, with almost a quarter (23.3 per cent) reporting regular shivering when in the open environment, compared with just 11 per cent of other households. This indicates that fuel-poor households, who are exposing themselves to sub-optimal levels of warmth indoors, appear to be exposing themselves to additional cold-related strains outdoors, compounding potential adverse health impacts. Such results are worrying when the age profile of the fuel-poor is considered; it is well known that the elderly suffer from fuel poverty more so than the young, and this is generally because reduced levels of disposable income associated with retirement make it more difficult to afford adequate home heating. Indeed, chronic fuel poverty is particularly high in Ireland, and to a lesser extent in the UK, among lone pensioners (Chapter 5). This finding may indicate that many older, more vulnerable individuals unable to heat their homes adequately may also be unable to clothe themselves adequately from the cold outdoors. However, it is also possible that such behaviour may not be income related, i.e. it may be a conscious decision not to wear heavy clothing, as opposed to an inability to afford adequate clothes.

Table 7.2 Fuel poverty and outdoor cold exposure

	Fuel poor % households	Other households %
Wrap up, rarely shiver	76.7	89.0
Cold, often shiver	23.3	11.0

Duration of Shivering Episodes Outdoors

If more fuel-poor households are enduring indoor *and* outdoor cold strain than other households, then it is useful to examine the duration of shivering spells both outside and inside the home and to identify associations with the fuel-poverty status of the household. Table 7.3 reports the results for the duration of shivering episodes outdoors.

Table 7.3 Fuel poverty and duration of outdoor cold exposure

	Fuel poor % households	Other households %
Not at all	38.0	57.4
1-10 mins.	44.4	34.0
10-30 mins.	12.8	6.5
30-60 mins.	3.2	1.4
>1 hour	1.6	0.6

The findings of the national survey demonstrate a clear relationship between fuel poverty and shivering spell outdoors with a far smaller proportion of fuel-poor households reporting no shivering at all when outdoors on a cold winter day (38 per cent) compared with other households (57.4 per cent). This difference of 19.6 percentage points represents a 51 per cent proportionate variation between the two population groups. Conversely, shivering spells, both short and long, are reported at repeatedly higher incidences by the fuel-poor. Some 44.4 per cent of the fuel-poor report shivering spells of up to 10 minutes on a typical, cold winter day, compared with 34 per cent of other households, a 31 per cent proportionate variation. Similar results are found for households that shiver typically for 10 to 30 minutes, with 12.8 per cent of fuel-poor households declaring shivering spells of this length, compared with 6.5 per cent of other households, a 97 per cent variation. Longer spells of shivering (over a half-hour and one hour or more) are reported at lower incidences among both household groups, but fuel-poor households remain over twice as likely to shiver for such durations than other households.

Duration of Shivering Episodes Indoors

The results of this question, relating to shivering episodes resulting from indoor cold exposure, are crucial in identifying the relationship between fuel poverty and inadequately heated homes. Households are asked to specify for how long they shiver from the cold when indoors on a cold winter night. Table 7.4 reports the results and shows a pattern of shivering similar to that identified with regard to outdoor shivering. However, the magnitude of the results is far greater in this section. Fuel-poor households are up to seven times more likely to shiver from the cold indoors than other (non-fuel-poor) households. Over half of fuel-poor households shiver indoors for up to a half an hour on a cold winter evening, compared with just 15.5 per cent of other households. In addition, a further 6 per cent of the fuel-poor shiver for longer periods, while other households report negligible levels of shivering for this duration of time. As might be expected, the vast majority (84.2 per cent) of households unaffected by fuel poverty do not shiver indoors at all during the winter. The most marked proportionate variation in duration of shivering spell occurs in the 10-30 minute category, with fuel-poor households reporting an incidence of shivering of some 6.7 times that reported among other households (16.1 per cent against 2.1 per cent). These results indicate that fuel-poor households are substantially more likely to be inadequately protected from the cold when indoors than other households, as might be expected.

Table 7.4 Fuel poverty and duration of indoor cold exposure

	Fuel poor % households	Other households %
Not at all	43.5	84.2
1-10 mins.	34.5	13.4
10-30 mins.	16.1	2.1
30 60 mins.	4.8	0.3
>1 hour	1.2	0

Lifestyle Risk Factors

As well as examining physiological responses to cold indoors and outdoors, a number of other questions relating to non-cold risk factors associated with ill health were asked in the national household survey of Ireland. The first set of these related to lifestyle risk factors pertaining to smoking habits, exercise levels and alcohol consumption. These results are reported in this section (Table 7.5). As well as being less well-protected from the cold indoors and outdoors, fuel-poor households appear to have higher smoking rates than other households. The national household survey calculated that 35.3 per cent of households in fuel poverty declare that they are 'regular' (i.e. daily) smokers, compared with 23.2 per cent of the rest of the population, a proportionate variation of 52 per cent. As smoking is considered an important risk factor associated with reduced health (Mannino et al., 2001, such a finding is notable. In addition, fuel-poor households are less likely to exercise regularly. When asked whether they engage in any regular (i.e. weekly) activity, such as jogging or cycling, long enough to work up a sweat, less than one-in-five fuel-poor households (19.4 per cent) responded affirmatively, compared to 28.9 per cent of other households. However, when asked about typical alcohol consumption, both household groups reported similar drinking patterns.

Table 7.5 Fuel poverty and lifestyle risk factors

	Fuel poor % households	Other households %
Regular smoker	35.3	23.2
Weekly exercise	19.4	28.9
Regular drinker	78.4	81.9

Diet

Diet is an important variable in any equation of health status (Forsyth et al., 1994). It is, therefore, useful to identify whether fuel-poor households eat less healthily than other households, thus exposing themselves to another exogenous risk factor associated will reduced health status. Table 7.6 illustrates the findings of the national survey and demonstrates some clear relationships.

Table 7.6 Fuel poverty and diet

	Fuel poor % households	Other households %
Fruit: < twice/week	32.2	18.9
Vegetables: < twice/week	7.0	3.1
Whole-wheat or rye bread: rarely or never	43.0	28.6
Sugary food: > once/day	16.2	10.0
Fried food: > twice/week	71.4	31.6

Five food-types are examined, two of which are considered undesirable and best eaten as little as possible (fried food and sugary food), three of which are considered highly desirable and nutritious and recommended by dieticians for a healthy diet (fruit, vegetables and whole-wheat or rye bread). Again, a clear pattern of dietary deprivation emerges from the survey which indicates that fuel-poor households eat far more unhealthily than other households. A third (32.2 per cent) of fuel-poor households eat less than two servings of fruit per week, compared with just 18.9 per cent of other households. Virtually all households unaffected by fuel poverty eat vegetables at least twice a week, while a small proportion (7 per cent) of the fuel-poor do not. Some 43 per cent of fuel-poor households 'rarely' or 'never' eat whole-wheat or rye bread, while the corresponding proportion for other households is found to be 28.6 per cent. With regard to unhealthy foodstuffs, sugary foods are eaten more than once a day by 16.2 per cent of fuel-poor households, compared with exactly one-in-ten of all other households. Furthermore, fried food is consumed more than twice a week by a remarkable 71.4 per cent of the fuel-poor, compared with just 31.6 per cent of other households, representing a proportionate variation of 126 per cent. Taken as a whole, these findings relating to dietary intake of households indicate that households unable to heat their homes adequately are also unable (or unwilling) to eat healthily.

Physical Activity

Levels of physical activity, in the form of regular exercise, were touched upon in an earlier section. This section explores physical activities further, by asking households about the degree of physical input required in their job as well as the level of light and heavy household work they perform on a weekly basis. Such data are useful in identifying broadly the overall levels of fitness of a population. A third of fuel-poor households reported that their job involved 'very' or 'fairly' high levels of physical activity when asked in the national household survey. However, almost a half of other households (47.5 per cent) responded similarly, indicating that proportionately fewer fuel-poor individuals are employed in jobs requiring moderate or high levels of physical activity. However, in the homeplace, similar levels of physical activity are found for both household groups. Approximately two-thirds of fuel-poor and non-suffering households state that they undertake light household work (e.g. dusting, washing dishes) at least three times weekly, while a third of both the fuel-poor and those unaffected by fuel poverty engage in heavy household work (e.g. washing floors and windows, vacuuming) more than three times per week. Overall, it would appear that the association between fuel poverty and the level of physical activity required in the workplace is a relatively more important risk factor than the degree of physical activities undertaken in the home.

Table 7.7 Fuel poverty and levels of physical activity

	Fuel poor % households	Other households %
Job: 'very' or 'fairly' physically active	33.7	47.5
Light household work: >3 times/week	68.5	63.9
Heavy household work: >3 times/week	34.2	34.0

Mode of Transport

Finally, the survey examined households' usual mode of transport when shopping. Fuel-poor households are less likely to drive by car to the shops; this is indicative of the financial position of those in fuel poverty being less buoyant than other households; two-fifths of fuel-poor households drive compared to three-fifths of other households. Instead, walking is far more frequently reported as a mode of transport by fuel-poor households, while there is also a similar proportionate variation in the incidence of public transport utilisation between the two groups. The use of a bicycle is negligible in both household groups. While the lower incidence of car usage and the higher incidence of walking among fuel-poor households may appear to be an apparently positive finding, the result is, in many ways, a double-edged sword. This is because walking to the shops entails exposure to the outdoor environment, and it was shown earlier that fuel-poor individuals are less likely to protect themselves adequately from the cold when outdoors, resulting in considerable episodes of shivering. Because many fuel-poor households are elderly, such exposure to the cold can be detrimental over time to the health of the individual, as has been shown in a variety of epidemiological studies (e.g. Eurowinter Group, 1997).

The 57 per cent proportionate increase in the number of fuel-poor households that use public transport over other households also has potential repercussions on human health. This is because waiting for a bus or train normally entails remaining relatively stationary for a time, and it is during this period of little physical activity that body temperature is susceptible to significant falls which can result in cardiovascular strain, especially among more elderly individuals. Therefore, while such results may ostensibly be interpreted as positive in terms of risk-factor analysis of health status, on closer examination such findings may well be more detrimental than beneficial, especially among a population comprised of relatively high numbers of older people.

Table 7.8 Fuel poverty and usual mode of transport

	Fuel poor % households	Other households %
Car	42.6	61.3
Walk	36.8	20.6
Bike	1.2	1.5
Public transport	10.5	6.7
N/A, other	8.9	10.0

Discussion

This research has employed results from a national household survey of Ireland, undertaken in 2001, to conduct a risk-factor analysis of fuel-poor households. There have been many studies highlighting the plight of the fuel-poor over the past decade in the UK and, to a lesser extent, in Ireland. Much of this research is useful, progressive and worthwhile work. This chapter does not argue the counter-factual, however it indicates that those wishing to tackle the problem of fuel poverty and related ill health are faced with a difficult, multifarious task. This is because it appears that fuel-poor households not only exhibit high levels of endogenous risk factors pertaining to indoor cold stress, but they also exhibit high levels of a range of other, exogenous, generalised risk factors associated with impaired health status. Therefore, on the basis of these data, it is difficult to argue that fuel poverty results in impaired health. However, fuel-poor households are more likely to suffer ill health than other households, but it is not just because they are fuel-poor. In this regard, policy-makers are faced with a multi-faceted job in mitigating fuel poverty-related ill health.

It was demonstrated that, while the duration of time spent outdoors during winter did not appear to exhibit a strong relationship with the fuel-poverty status of a household, the presence and duration of shivering outdoors was a more telling variable. Such a physiological outcome measure indicates inadequate thermal protection from the cold, and fuel-poor households were shown to suffer more, and for longer periods of time, from spells of cold-induced shivering. The results were dramatic when the question was replicated to elicit information regarding households' indoor environment. Some 56.5 per cent of fuel-poor households suffer from episodes of shivering indoors on a typical, cold winter night, compared with just 15.8 per cent of those unafflicted by fuel poverty. Such findings are very similar to those found in previous epidemiological studies on the adverse health effects of cold stress. The Eurowinter Group (1997) argued that, while cold housing already receives much, well-earned attention, there should also be scope for action to reduce mortality from outdoor cold exposure. Donaldson et al. (1998) concurred and demonstrated that excess winter mortality could be reduced in western Europe by people wearing sufficient clothing and engaging in physical activity outdoors, while adequately heating their homes indoors. The relative importance of indoor and outdoor cold exposure on health are difficult to disentangle precisely, but the findings of this research and the work of Donaldson et al., in particular, indicate that both play key roles in seasonal variations in morbidity and mortality in western Europe.

However, when the risk-factor analysis examined other, exogenous (non-cold) hazards associated with impaired health, fuel-poor households were found to fare worse than other households. Specifically, fuel-poor households smoke more, exercise less and eat less healthily than other households. These findings suggest that socio-economics also play a large part in health inequalities associated with the presence of fuel poverty. This is not a radical result; Kunst et al. (1991) showed that the reduction in excess winter deaths in the Netherlands was closely related to factors associated with socio-economic progress, and Townsend has always stressed the importance of the class structure in his materialist explanation of health divides.

Current strategies aimed at combating fuel poverty in the UK and in Ireland are based on retrofitting the homes of low-income households with energy-saving measures to improve the energy efficiency of the dwelling. Such programmes are highly economically desirable, resulting in large energy cost savings and attendant external benefits such as reductions in environmental emissions, and other health and comfort benefits accruing to both the exchequer and the individual (Clinch and Healy, 2001). Notwithstanding this, it is logical to conclude from this study that such programmes are unlikely to result in the alleviation of all ill health related to fuel poverty, as it appears from this study that fuel-poor households, like those living in generalised income poverty and deprivation, suffer from a plateau of other health hazards associated with being less well-off.

It is clear from previous research that fuel-poor households are more likely to exhibit low levels of educational attainment (Chapter 6). Therefore, it is important to maintain and improve current health-promotion strategies informing individuals about the risks attached to smoking, the importance to eat as healthily as possible and to exercise regularly. On the other hand, it is also important to pursue vigorous information and awareness campaigns regarding the importance of domestic energy efficiency as a means to achieving affordable warmth in the home. There is a remarkable 'information gap' regarding the benefits of domestic energy efficiency; a recent national household study in Ireland showed that, when asked why they had not retrofitted energy-saving measures into their home, over half of all respondents replied that they were either unaware of such measures or they did not know of the benefits of installing such technologies (Chapter 6). Reducing the information gap should entice more affluent households to retrofit their homes using personal funds, however it is clear that the more economically vulnerable require some form of state assistance in carrying out such remedial work. This should continue, but only in tandem with other health-promotion strategies. There is very little macro awareness, at least in Ireland, of the dangers associated with inadequate protection from the cold outdoors. It is suggested in this chapter that policy-makers give this area some thought, as the results of this (and other research) demonstrate that cardiovascular strain resulting from outdoor cold exposure is an important risk factor associated with adverse health conditions, most especially among the elderly in society.

In short, this chapter argues that programmes aimed at alleviating fuel poverty, while beneficial as part of a package of measures, will not succeed on their own in eradicating all fuel poverty-related health impairment. Government-backed strategies must also incorporate information components regarding the hazards of inadequate protection from the cold outdoors (and indoors), as well as more general health-promotion campaigns regarding the hazards of lifestyle risk factors such as smoking and unhealthy diets.

Chapter 8

Fuel Poverty, Thermal Comfort and Household Occupancy in Ireland

Introduction

It is now well documented that fuel poverty has a number of adverse health impacts, especially on the elderly. Chronic exposure to low ambient temperatures in the home resulting from fuel poverty often leads to a physiological condition in humans known as cold strain. While short episodes of cold stress are unlikely to cause serious adverse health impacts among the young and healthy, such physiological effects are damaging to the cardiovascular and respiratory system of the elderly, and may exacerbate current ill health or diminish resistance to infections in healthy persons (Collins, 1986). At worst, chronic cold strain can result in fatal conditions and premature mortality. It is, perhaps, unsurprising that Ireland and the UK have the highest levels of seasonal mortality in Europe (Clinch and Healy, 1999a). Much of these excess winter deaths have been associated with inadequate protection from indoor cold stress and fuel poverty more generally (Clinch and Healy, 2000c, Eurowinter Group, 1997).[1] In addition, research presented in Chapter 7 has shown that fuel-poor households persistently report lower levels of health status using objective and subjective outcome measures. The variation in health appears to be most acute for chronic diseases of the respiratory system and long-term depression.

Besides issues relating to the adverse health effects of fuel poverty, there are other concerns regarding thermal comfort that have not been addressed satisfactorily. There has been virtually no work examining the relationship between fuel poverty and the extent of indoor cold strain; such data would be particularly useful for the elderly, where the strongest health and comfort impacts are likely to arise. The thermal comfort of housing has not been assessed empirically in Ireland hitherto, and there have been relatively few such studies in the UK. The effect of fuel poverty on thermal comfort and household temperature is also in need of more exploration. In addition, the effects of fuel poverty on household occupancy have not been addressed formerly using empirical analysis. This analysis contributes to the literature on fuel poverty and comfort by addressing these key issues. The main hypotheses explored in this chapter are:

---

[1] Although there is evidence demonstrating that fuel-poor households also exhibit higher lifestyle risk factors associated with impaired health, as can be seen in Chapter 9.

- Fuel-poor households live in colder homes
- Fuel-poor households endure higher levels of thermal discomfort
- Fuel poverty results in adverse occupancy effects

In this chapter, the Urban Institute Ireland national household survey of Ireland was again employed. Households were asked a variety of questions relating to the thermal comfort of their home on a room-by-room basis to assess the magnitude of the relationship between fuel poverty and comfort. These self-reporting data act as good subjective measurements of thermal comfort. Living room temperature readings were taken in all 1,500 surveyed households so that the thermal comfort of fuel-poor households could be compared with other households using an objective measurement. Households are also asked about the extent and duration of indoor cold strain during winter. The level of shivering (a basic physiological response to cold strain) resulting from indoor cold stress is compared among the fuel-poor and other households. The chapter then turns to occupancy issues relating to fuel poverty. Those who occupy the home the most, and are therefore most at risk of the health and comfort effects of fuel poverty, are identified, as are the incidences of fuel poverty across various household typologies. In addition, households are asked to consider whether rooms are occupied in the wintertime even if they are cold and unheated. Conversely, households are also assessed to evaluate whether certain rooms are unoccupied because they are too cold.

National Household Survey

The Data

The national household survey, conducted in 2001, is employed in this chapter. Results are only reported in this study where sample sizes allow for a reasonably low level of the margin of error. When the whole sample is being considered and comparisons are being made between two groups, the margin of error using a 95% confidence interval is +/-2.5%, i.e. differences of more than 5% can be said to be 'significant' with 95% confidence. Where frequency counts are too small to allow for reliable comparisons, a dash, denoted '-', appears in the data tables. The chapter does not devote space to detailed statistical or econometric analyses of the results.[2] Rather, for the sake of clarity and brevity, the chapter presents key findings relating to subjective and objective measures of thermal comfort, and the relationship between such measures and fuel poverty. The impact of occupancy on fuel poverty is also examined, and special emphasis is placed on results pertaining to the elderly population.

---

[2] A detailed, multivariate, Probit regression analysis of fuel poverty can be found in in Chapters 5 and 6.

Defining Fuel Poverty

The definition of fuel poverty employed in this study is the standard qualitative definition first used by Lewis (1982) and later modified by Clinch and Healy (1999a) 'the inability to heat the home adequately because of low household income and energy inefficient housing'.

Households in the survey are first split into two discrete groups: fuel-poor households (those declaring an inability to adequately heat the home to a comfortable temperature) and other (non-fuel-poor) households. Some 17.4% of the sample (representing 226,000 households) report some level of fuel poverty in Ireland. It was found that over a quarter of these fuel-poor households (27%) suffer from chronic fuel poverty, i.e. they are persistently unable to afford to heat their homes adequately over time. The remaining fuel-poverty sufferers suffer more occasional difficulties in achieving affordable warmth. While the incidence of chronic fuel poverty appears to have declined in Ireland by over two-fifths since 1994, the overall level remains very high relative to other European countries.

It is also likely that much of the reduction in the incidence of fuel poverty over the period 1994-2001 is owing to macroeconomic variables. Ireland has had spectacular economic growth since 1994 (with increases in GDP per capita averaging over 8% per annum during this time period), and it is thought that many low-income households have been pulled out of the (fuel-) poverty trap in this time. However, it is likely that a serious downturn in the economy would result in substantial increases in the numbers affected once more.

Defining Thermal Comfort

Many physiological, psychological and environmental variables play a part in humans' perception of thermal comfort. The most important physical parameters include air temperature, air velocity, relative humidity and the mean radiant temperature of surrounding surfaces (Fanger, 1972). However, it is possible to identify an objective level of thermal comfort using indoor temperature as a gauge. Humans require an environment which is of a temperature that poses no additional health risks. If 'health' is taken to mean normal physiological functioning in the absence of stress, such as that produced by thermal discomfort, then the temperature range 18 24°C poses little threat to sedentary, healthy people adequately clothed (Collins, 1986). The Building Research Establishment has generally considered 18-21°C as a comfortable temperature for a living room during wintertime (BRE, 1995), and Boardman has generally advocated this temperature as 'comfortable' (Boardman, 1991). In addition, the UK Government employs 21°C as an adequate level of warmth for the living room, while 18°C or more is acceptable for all other areas (DEFRA, 1999).

However, 18°C is the benchmark set by the World Health Organisation (Collins, 1986), with increases of 2-3°C for those more vulnerable to the effects of cold strain, such as the sedentary elderly, the young and the physically disabled. A temperature of 18°C has also been used as a baseline comfortable temperature in other studies (Mant and Muir-Gray, 1986; Raw and Hamilton, 1995). This study employs the WHO's (lower bound) temperature benchmark of 18°C as the basic temperature level required for thermal comfort for the non-elderly population. For those aged 65 or more, thermal comfort is defined in the temperature range 20-24°C, in line with the WHO's guidelines. The adoption of the WHO's 'comfort zone' ensures more reliable, lower bound results in the study.

Results: Thermal Comfort

Over the next two key sections of the chapter, the results of the national household survey of Ireland pertaining to thermal comfort and occupancy (and their relationship with the fuel-poverty status of the household) are presented.

Thermal Comfort and Fuel Poverty

Households are asked to rate the level of comfort, in terms of temperature, in each room of their house. The response variable was a seven-point scale which ranged from 'hot' to 'cold' with the neutral point denoting 'comfortable' or representing the self-perceived correct temperature. All households reporting levels of thermal comfort that deviate from the 'comfortable' response outcome can, therefore, be considered as enduring sub-optimal levels of thermal comfort. The results demonstrate that fuel-poor households persistently endure poorer levels of thermal comfort than other households, with each room in a fuel-poor household reported as less thermally comfortable.

Over a quarter of fuel-poor households (25.6%) report either too hot or too cold living rooms, while just 9.1% of other households declare such problems, representing a proportionate variation of 181%.[3] Large variations are found among households regarding comfort levels in kitchens. Almost a third of fuel-poor households (compared with 13.6% of other households) demonstrate sub-optimal comfort conditions in this room. The results for the dining room or study are not quite as dramatic, with 15.8% of fuel-poor households reporting uncomfortable temperatures compared with 8.4% of other households. The master bedroom is occupied, on average, for over 8 hours per day and, as such, results regarding the thermal comfort of this room may be considered relatively important. It is, therefore, interesting that the most stark variations in thermal comfort are found in this room. Just 6.7% of households unaffected by fuel poverty report uncomfortable temperatures in the master bedroom, compared to 32.6% of fuel-poor households. Results for other bedrooms are similarly stark. Some 37.6% of fuel-poor households report poor comfort levels in their second bedroom, compared with just 9% of other households, while some two-in-five fuel-poor

---

[3] It is useful to include households that are suffering thermal discomfort because of overly hot household temperatures. While it is unlikely that many fuel-poor households are suffering from this state, it is possible that inadequate heating controls and inadequate heating facilities force some fuel-poor households into a situation where sub-optimal hot temperatures are attained.

households declare poor thermal comfort in their third bedroom, compared to 10.9% of other households. The highest overall incidence of sub-optimal levels of thermal comfort is found in the bathroom, where 44.6% of fuel-poor households report comfort deficits, compared to 9.2% of all other households.

The results demonstrate that not all those who are classified as fuel-poor self-report thermal discomfort in the home. This implies that some households who report an inability to afford to heat the home adequately (i.e. the fuel-poor) may be spending beyond their means in an attempt to achieve adequate household warmth in the winter. It is also likely that the data 'under-declare' the levels of thermal discomfort in the home, as is often the case with self-reported data.[4] However, the most plausible reason for this is that many of those classified in this study as 'fuel-poor' are suffering from intermittent strains of fuel poverty (as was reported in Table 6.2). Such fuel-poor households are less likely to report thermal discomfort than chronic sufferers who account for a smaller proportion of total fuel-poor households.

When the data are disaggregated to examine elderly households only, the results show that, for each room, the levels of thermal discomfort are higher among the over-65s than for other households. Over a fifth of the elderly declare thermal discomfort in the bathroom. Over a quarter (28.8%) demonstrate thermal discomfort in their third bedroom, 24% in their second bedroom and 13.1% in the master bedroom. Just over one-in-eight households over 65 years of age report thermal discomfort in the living room. Such results show increased levels of sub-optimal thermal comfort among the elderly population which is cause for concern when it is considered that such households are far more vulnerable to the adverse health effects of low ambient temperature in the home than other households.

Table 8.1 Fuel poverty and % of households with sub-optimal thermal comfort

	Fuel-poor households %	Other households %	Households >65 years' old %
Living room	25.6	9.1	12.9
Dining room/study	15.8	8.4	17.1
Kitchen	30.6	13.6	14.4
Master bedroom	32.6	6.7	13.1
2nd bedroom	37.6	9.0	24.0
3rd bedroom	40.0	10.9	28.8
Other bedrooms	25.0	11.1	16.1
Bathroom	44.6	9.2	20.5

---

[4] It was shown in the cross-European analysis of housing conditions, presented in Chapter 4, and in the results pertaining to fuel poverty and health (Chapter 7) that Irish and Italian households persistently 'under-declare' their respective levels of hardship and housing deprivation, especially when in relation to health impairment.

Living Room Temperature and FuelPpoverty

The temperature of the living room is recorded in the national household survey. The thermometer is placed in the main living room, at table height, at least one metre away from any person or source of heat. Temperatures of between 12°C and 26°C were recorded across all households during March 2001.[5] This outcome measure acts as a good objective measurement of thermal comfort in the household, although it is recognised that living-room temperature is by no means a flawless gauge of thermal comfort.[6] However, the results, presented in Table 8.3, demonstrate that those households enduring fuel poverty are more likely to be living in colder homes, as might be expected, exposing themselves to increased risks of impaired health status. Some 29.4% of fuel-poor households demonstrated a living-room temperature of 18°C or less, compared with just 8.8% of other households.

Such results are startling when it is considered that resistance to respiratory infections has been shown to be diminished at such temperatures. A remarkable 68.6% of fuel-poor households are found to have temperatures in their living room of less than 20°C, while the corresponding proportion for all other (non-fuel-poor) households is 49.3%. Such a temperature is still lower than that frequently specified as thermally comfortable for a living room by the Building Research Establishment (BRE, 1995), Boardman (Boardman, 1991) and the UK Government (DEFRA, 1999). Therefore, about one-in-seven Irish households, or 14%, and almost three-in-ten fuel-poor households contain living rooms where the temperature is below that recommended by the World Health Organisation as a minimum temperature required to avoid cold strain and related ill health (Collins, 1986). In addition, over two-thirds of fuel-poor households and almost half of all other households in Ireland are failing to achieve the 'satisfactory heating regime' specified by the UK government. This implies that either the 'satisfactory heating regime' delineated by the UK government is unfeasible and of too high a standard, or that large portions of households in Ireland are living in unsatisfactorily heated homes.

It is somewhat disturbing to examine the temperature readings by age of the householder. Households consisting of elderly persons, who generally require a minimum temperature of at least 20°C to avoid cold strain, are shown to be less well-heated than other households. A remarkable 50.7% of such households fail to achieve the absolute minimum living-room temperature of 20°C set by the World Health Organisation for the elderly. Moreover, 16.1% of households over 65 years of age contain living rooms with temperatures of less than 18°C, while even a small number were found to have temperatures below (some considerably below) 16°C. On the other hand, 5.4% of households containing persons aged over 65

---

[5] Outdoor temperature in March was about the March average (6.3°C) for the duration of the temperature measurements undertaken in the national household survey.

[6] Households may, in anticipation of the interview, heat the room to a higher level than that to which it is normally heated. In addition, a warm living room can be found in an otherwise cold house.

were shown to have temperatures *exceeding* those specified within the WHO 'comfort zone' for the elderly of 20-24°C.

Table 8.2 Fuel poverty and living room temperature

	Fuel-poor households %	Other households %	Households >65 years' old %
<16°C	5.5	1.8	1.5
16-17.9°C	23.9	9.0	14.6
18-19.9°C	39.2	38.5	34.6
20-21.9°C	19.2	33.1	32.7
22-23.9°C	9.0	11.4	11.2
24-25.9°C	3.1	6.0	5.4
≥26°C	-	-	-

The results demonstrate that not all those who are defined in the study as fuel-poor can be classified as enduring thermal discomfort, i.e. there is a mismatch between the two classifications. This highlights the problems associated with using living-room temperature as a measure of thermal comfort. It has been found that living-room temperature is often not a good indication of whole-house temperature in the UK (Milne and Boardman, 2000), and it is likely that this finding applies to Ireland also. However, it is likely that the data are also demonstrating the phenomenon of intermittent or occasional fuel poverty, as opposed to chronic fuel poverty. It has been shown that the former type of fuel poverty is, by far, the most frequently found in Ireland, and it is likely that those households that are defined as fuel-poor in the study but who demonstrate adequate ambient temperature in the home are suffering from this intermittent level of fuel poverty, reported in Table 6.2.

When the results are disaggregated by social class it appears that the poor, as might be expected, are less likely to achieve a living-room temperature of 18°C. For instance, one-in-five of those in social class D (semi-skilled) demonstrate cold living-room temperatures of below 18°C, compared with about one-in-ten households in social classes A (professional) and B (managerial and professional). Households with large numbers of dependent children (four or more) exhibit the highest incidences of cold living-room temperature, with 16% of such households affected; perhaps more worryingly, the majority (62.1%) of such households contain living rooms of less than 20°C which is the minimum temperature recommended for young children and infants (Collins, 1986).

Indoor Cold Strain and Fuel Poverty

This section analyses the extent and duration of self-reported indoor cold strain on a typical cold winter night. The data are examined to assess whether fuel-poor households are more or less likely to endure indoor cold strain than other households. As an outcome gauge, shivering is a useful physiological measure which indicates that the body is not adequately protected from a cold environment and is experiencing some level of thermoregulation. While short spells of shivering are relatively harmless among the young, such physiological reflexes are thought to be more pernicious to older individuals, leading to cardiovascular strain and potentially other, more severe manifestations in the form of adverse health conditions such as hypothermia and pneumonia (Collins, 1986; Collins and Exton-smith, 1983).

The results of this question, relating to shivering episodes resulting from indoor cold exposure, are useful in identifying the relationship between fuel poverty and inadequately heated homes. Households are asked to specify for how long they shiver from the cold when indoors on a typical cold winter night. Table 8.3 reports the results and shows high incidences of cold strain among fuel-poor households; in fact, those living in fuel poverty are up to seven times more likely to shiver from the cold indoors than other households. Over half of fuel-poor households shiver indoors for up to half an hour on a cold winter evening, compared with just 15.5% of other households. In addition, a further 6% of the fuel-poor shiver for longer periods, while other households report negligible levels of shivering for this duration of time. As might be expected, the vast majority (84.2%) of households unaffected by fuel poverty do not shiver indoors at all during the winter. The most marked proportionate variation in duration of shivering spell occurs in the 10-30 minute category, with fuel-poor households reporting an incidence of shivering of some 6.7 times that reported among other households (16.1% against 2.1%). These results indicate that fuel-poor households are substantially more likely to be inadequately protected from the cold when indoors than other households, as might be expected.

Table 8.3 Fuel poverty and duration of indoor cold strain

	Fuel poor % households	Other households %	Households >65 years' old %
Not at all	43.5	84.2	70.2
1-10 mins.	34.5	13.4	22.5
11-30 mins.	16.1	2.1	5.5
31-60 mins.	4.8	0.3	1.4
>1 hour	1.2	0	-

In addition to the strong relationship found between the presence of fuel poverty and indoor cold strain, the data show that a substantial proportion of elderly households are suffering from episodes of shivering. Approximately 30% of those over 65 years' old report shivering from the cold indoors for various durations during the winter. While a relatively small number of these households demonstrate cold strain for extensive periods of more than half an hour, the results are still of strong significance. It has been understood for some time now that even relatively short periods of cold strain can result in adverse health impacts on the elderly, especially those sedentary in the home (Collins and Exton-Smith, 1983). In this regard, the finding that almost a third of elderly households expose themselves to cold strain during a typical winter night is cause for concern.

Results: Occupancy

There has been very little empirical work undertaken on the effects of household occupancy on fuel poverty. The national household survey asked some questions in this respect to elicit information regarding who occupies the house most and whether fuel poverty results in a differing household occupancy patterns.

Who Occupies the House Most and Fuel Poverty

Married women under 65 years of age occupy the home the most across all households, where 62.8% of households are mostly occupied by such householders, and among fuel-poor households (44.5%). However, it appears from the data that, among men, those whose marital status is single and who are aged under 65 are more likely to occupy fuel-poor households most of the time (11% compared with 6%), as are single women under 65 (13.8% versus 8.7%). One-in-ten fuel-poor households are occupied mostly by married women over 65, while married men aged over 65 are found in about half as many fuel-poor households.

Table 8.4 Fuel poverty and who occupies the house most

	Fuel poor % households	Other households %	Incidence of fuel poverty %
Married woman 65+	10.2	8.4	20.2
Married woman <65	44.5	62.8	12.5
Married man 65+	5.5	3.3	25.5
Married man <65	5.8	5.7	17.6
Child/ren	1.5	1.1	23.5
Single male <65	11.0	6.0	27.5
Single female <65	13.8	8.7	24.6
Single male 65+	4.7	2.8	26.1
Single female 65+	3.2	1.2	34.8

The incidence of fuel poverty by household occupancy is also calculated. It is found that the elderly are, in general, suffering from the highest incidences of fuel poverty. This is most acute among unmarried elderly households. The highest incidence of fuel poverty is found among lone female pensioner groups (over 65 years of age) where over a third (34.8%) are reporting fuel poverty. The corresponding incidence of fuel poverty for male lone pensioners is also relatively high, at 26.1%. The effect of marriage is positive in terms of reducing the incidence of fuel poverty, especially among women, although both married men and women over 65 also face high levels of hardship relative to younger people. It is found that 20.2% of married females over 65 and 25.5% of married males over 65 are affected. High incidences of fuel poverty were also found among single men and women aged under-65 years (27.5% and 24.6% respectively). Thus, the sex of the person who occupies the house most is also a factor in the incidence of fuel poverty.

Manifestations of Fuel Poverty on Occupancy and the Elderly

Households are asked about whether there are rooms in their home which remain unoccupied because they are unheated and are, thus, too cold to inhabit. Furthermore, households are asked whether rooms that are unheated and cold are occupied for at least ten minutes of the day. The results of the national household survey demonstrate that one-third of fuel-poor households often occupy cold, unheated rooms, compared to just one-in-ten of other households. Some 17.6% of elderly households occupy cold rooms for a time. While fuel-poor households occupy cold rooms more so than other households, they do not leave rooms unoccupied because they are too cold, neither do all other households. In fact, the proportion of households who do not inhabit cold rooms across Ireland is negligible, indicating that households would, in general, rather live in an uncrowded environment than a cold one. Nonetheless, one-fifth of elderly households declare that rooms are unoccupied because they are unheated and too cold.

Table 8.5 Fuel poverty and manifestations on occupancy

	Fuel poor % households	Other households %	Households >65 years' old %
Cold, unheated rooms occupied	32.3	10.4	17.6
Rooms unoccupied because too cold	3.5	2.2	19.4

Discussion[7]

The results of the national household survey indicate that there are large numbers of fuel-poor households living in thermal discomfort. Self-reported measures of thermal comfort demonstrate that about one-in-ten households (which represents 130,000 homes) nationwide report thermal discomfort, with higher levels in kitchens and bedrooms, while fuel-poor households are up to five times more likely to report thermal discomfort in certain rooms in their homes, such as the master bedroom and the bathroom. Furthermore, over half of fuel-poor households endure shivering episodes on typical cold winter evenings; such physiological

---

[7] More discussion on the policy implications of the findings in this chapter can be found in Healy and Clinch (2002).

responses to cold temperatures indicate cold strain, reduced resistance to respiratory infections and potential cardiovascular stress. When more objective output measures are employed, a similar pattern of thermal discomfort is demonstrated, with those enduring fuel poverty far more likely to be inhabiting homes with living-room temperatures below those set by the World Health Organisation as minimum levels of warmth to sustain resistance to impaired health status. Overall, three-in-ten fuel-poor households live in dwellings where the living room is heated to levels which can result in adverse health impacts, even on the young and healthy. Not all fuel-poor households are demonstrating thermal discomfort. It appears from the survey that many fuel-poverty sufferers in Ireland demonstrate occasional difficulties in achieving affordable home heating, while chronic sufferers appear to endure thermal discomfort and the coldest housing.

The age profile of those living in thermally uncomfortable housing is telling. Substantial portions of the over-65s are enduring cold housing. Using self-reported data, over a quarter of those over 65 are declaring thermal discomfort in some rooms in their home. Furthermore, about 30% of the elderly report episodic shivering which may result in cardiovascular strain in older people. However, the results based on the objective data on ambient living-room temperature make for more alarming reading, with over a half of old-age pensioners living in housing where temperatures are below those set by the WHO as minimum satisfactory levels of warmth for the vulnerable. A similar age profile of the fuel-poor is also found, with high incidences reported among those over 65, especially those unmarried – single female-pensioner households demonstrate an incidence of fuel poverty of 34.8%. Socio-economic analysis demonstrates that those in poorer social groups are far more likely to live in cold housing, as are those with large numbers of dependent children. Such a finding is worrying when it is considered that young children generally require warmer ambient conditions than healthy adults. Thus, many young children may be living in under-heated homes, exposing themselves at a young age to health risks which may reduce resistance to infection.

The existence of fuel poverty also plays a part in household occupancy. One-third of fuel-poor households inhabit cold, unheated rooms during winter. In addition, a small proportion of households nationally do not use rooms because they are too cold to inhabit during the winter. Approximately one-fifth of households occupied by older people follow these occupancy patterns related to fuel poverty.

These results suggest that there are significant comfort benefits which would accrue from improving the thermal standards of the housing stock. A recent, *ex ante* study in Ireland concluded that the private monetary benefits to households in terms of increased thermal comfort could be as much as €461m (Clinch and Healy, 2002).

Chapter 9

Excess Winter Mortality in the EU: Identifying Key Risk Factors

Introduction

Excess winter mortality has been documented in medical journals for about 150 years (Guy, 1858), and most countries suffer from 5-30% excess winter mortality (Curwen, 1991). However, there still remains much debate with regard to why certain countries experience dramatically higher rates of seasonal mortality than others. Cold strain from both indoors and outdoors has been implicated on several occasions (Clinch and Healy, 2000c, Eng and Mercer, 1998; Eurowinter Group, 1997), however other potential factors (other than cold strain) have rarely been analysed. In addition, there has been virtually no published research on seasonal variations in mortality in southern Europe. This may be due to the perception that such countries are not affected by excess winter deaths because of their mild winter climates. This analysis demonstrates that such a perception is highly mistaken.

The overall health of a population is influenced by a large number of factors which can be grouped into four main categories. Besides factors associated with biological and genetic considerations which have been linked with reduced health status (Goldberg, 1963), there are three other categories to which most of the health divide apparent in modern industrial societies can be attributed (Townsend, 1979). First, environmental factors (social, economic and natural) play key roles in explaining health inequalities. It has been shown by many studies that mortality rates are negatively associated with a country's macroeconomic health, i.e. rates of age-adjusted mortality fall as national income increases (Cumper, 1981; Brainerd, 1998). State expenditure on education has also been shown to be associated with the health and well-being of many populations (Muller, 2002; Backlund *et al.*, 1999). If the socio-economic level of development, as measured by per-capita GDP or public expenditure on education, is an important predictor of the health (and, therefore, mortality rates) of a population, then it is thought beneficial to see if this relationship holds for excess winter mortality.

The second group of factors associated with the health status of a population relates to healthcare provision and health expenditure in absolute and relative terms. Both variables have been found to be negatively associated with all-year mortality rates in a considerably varied literature (Elola *et al.*, 1995). This study examines a variety of indicators of healthcare provision to identify whether cross-country variations in excess winter mortality are correlated with variations in standards of healthcare services in Europe.

Finally, environment and lifestyles are both strongly linked with a population's health (Poikolainen and Eskola, 1988). There is a very large literature analysing lifestyle risk factors with non-seasonal mortality. Smoking rates have been demonstrated to be highly associated with impaired health and premature mortality (Kondo et al., 1997; Boyle and Maisonneuve, 1995), as has obesity (Elia, 2001; Visscher *et al.*, 2001), but virtually none examine the potential influence of such factors on seasonal variations in mortality. Furthermore, there is a considerable epidemiological literature showing various degrees of (positive) relationships between all-year mortality and socio-economic indicators, such as income poverty (Shibuya *et al.*, 2002; Wildman, 2001), inequality (Kennedy *et al.*, 1996; Wolfson *et al.*, 1999) and deprivation (Wilkinson, 1996; Deaton, 2001a, 2001b). A smaller medical literature is evident regarding the association between fuel poverty (and poor domestic thermal efficiency) and excess winter mortality (Clinch and Healy, 2000c; Eng and Mercer, 1998; Eurowinter Group, 1997; Gemmell *et al.*, 2000). The relative importance of all of these factors on seasonal variations in mortality has not been explored hitherto. The basis of this chapter is to present a macro analysis of the broad relationship between these factors and excess winter mortality by employing, for the first time, a pan-European analysis. This is achieved by examining longitudinal winter mortality in 14 European countries. Multiple time-series data on a variety of risk factors are analysed against seasonal-mortality patterns to identify associations.

Methods

Data on mortality was first obtained. The primary source for the original data was the United Nations Databank.[1] This resource contains monthly mortality figures for many countries; however, some countries fail to supply up-to-date statistics, with the result that some data-sets were either somewhat or very incomplete.[2] Thus, the next step was to make individual contact with the chief statistical institute in each country for which there was incomplete information in the UN databank. This involved contacting the following countries: Denmark, England, Finland, France, Germany, Greece, Italy, Luxembourg, the Netherlands, Northern Ireland, Norway, Portugal, Scotland, Spain, Sweden and Wales. The data obtained was deaths from 'all causes' (International Classification of Diseases (ICD) 000-999).

Data on crude mortality, age-adjusted mortality and life expectancy were obtained from the World Bank's databank.[3] These data were crucial for comparing the demographic structures of various nations and for subsequent analysis detailed later in the chapter.

---

[1] I am deeply grateful to Seiffe Tadesse of the UN Databank for supplying me with this invaluable data.
[2] Personal communication with Seiffe Tadesse (Tadesse, 2000) confirmed that many countries do not respond to the UN questionnaires each year, leading to incomplete datasets.
[3] I am particularly grateful to Anat Lewin of the World Bank for furnishing me with these (and many other) crucial datasets.

A recent ten-year time-series (1988-97) was chosen for the baseline analysis; data for years post-1997 were not available for all 14 countries. Excess winter mortality is defined as the surplus number of deaths occurring during the winter season (December to March inclusive) compared to the average of the non-winter seasons; adjustments were made for leap years. A relative definition of seasonal mortality is employed in this analysis (Laake and Sverre, 1996), which enables meaningful cross-country comparisons of excess winter mortality to be achieved. The coefficient of seasonal variation (CSVM) in mortality is calculated using the following formula which acts as a lower bound estimate of seasonal mortality:

$$\text{CSVM} = [f_{deaths}(Dec+Jan+Feb+Mar)] - [f_{deaths}(Apr+May+Jun+Jul) + f_{deaths}(Aug+Sep+Oct+Nov)/2]$$

all divided by

$$[f_{deaths}(Apr+May+Jun+Jul) + f_{deaths}(Aug+Sep+Oct+Nov)/2]$$

All 14 European countries employed in the analysis, being relatively alike in economic and social characteristics, exhibit similar crude mortality rates (9-11 deaths per thousand population). Monthly mortality data were originally obtained through the United Nations Databank. Individual countries were contacted subsequently to obtain missing monthly mortality data.

A Poisson regression model is considered most suitable for the data, and it is employed on each risk factor to identify associations with seasonality over time. Mean monthly climatic data (precipitation, relative humidity and environmental temperature) were obtained using the Meteotest *Meteonorm V.4.0* CD-ROM, a meteorological computer program which contains reliable 30-year averages for several hundred weather stations globally. Weather stations were selected carefully on the basis that they were most climatically representative of a given country with regard to respective population dispersals. The weather stations employed were often those found in a country's capital city, as these captured, for the majority of cases, the largest share of the country's population. However, for those countries with particularly dispersed populations and discernible climatic variations (Germany, Italy, France), a north-south gradient was used, i.e. the average of two weather stations – each climatically representative and with high population densities – were employed.

In terms of population densities, Belgium is remarkable for two key reasons: it is a country with a surprisingly dispersed population, and Brussels (the capital city) is not the most populated city – it is in fact the fifth.[4] However,

---

[4] It is perhaps surprising that Antwerp (population: 465,000), Ghent (population: 230,000), Charleroi (population: 207,000) and Liège (population: 197,000) are all more densely

climatic variations are not especially dramatic across Belgium and so weather data from Uccle (close to Brussels) is used in the analysis.

A north-south gradient was employed for France, Germany and Italy. This is because of these countries' large population dispersal and very considerable climatic variations from north to south. Paris (the capital city) and Marseilles were chosen as France's representative northern and southern weather stations; they are also the two largest cities in France.[5] In Germany, Berlin (the capital city) and Munich were selected as representative north-south weather stations.[6] The two largest Italian cities, Rome and Milan, act as very good north-south gradients demonstrating the considerably varied climatic conditions found in Italy.[7] Scottish data is taken from Glasgow (population: 616,000), the largest city in Scotland, larger than the capital, Edinburgh (population: 449,000).

Longitudinal datasets on macroeconomic indicators were obtained from the United Nations Statistics Division and the World Bank. Time-series datasets were also obtained from the World Bank regarding lifestyle risk factors such as smoking and obesity, and on health-service provision. Data on four socio-economic variables were calculated using the European Community Household Panel longitudinal users' database covering the four years 1994-97; this survey is the first comparable, cross-country database on social indicators in the EU. As it only commenced in 1994, there are no cross-country data available in Europe regarding such socio-economic indicators prior to this year. In this regard, relative excess winter mortality is re-calculated for all 14 countries in the model for the socio-economic section of the analysis using a comparable time-series (1994-97). Income poverty is calculated for these years by assigning a poverty threshold of 60% of median equivalised income adjusted for purchasing power, as is often employed in cross-country poverty studies (Duncan *et al.*, 1993; Callan *et al.*, 1993). Income inequality is calculated using the Gini-coefficient measure. The Gini coefficient is calculated in a number of steps for each Wave of the ECHP results. The first step involves deciding on the variable for which the Gini-coefficient is to be estimated. It is standard practise to consider net (as opposed to gross) household income adjusted for household size (Eurostat, 1999). Therefore, total net household income is divided by equivalised household size according to the OECD modified scale. Households then have to be sorted according to adjusted net household income (sorting order: lowest to highest value). Based on the household files of the ECHP Users' Database, the Gini-coefficient can be calculated as follows:

---

populated than Brussels (population: 134,000), though Greater Brussels is itself the largest urban agglomeration in Europe.

[5] The Parisian population stands at 2,152,000, while Marseille – the next largest city – has a population of 874,000.

[6] Berlin's population is currently 3,472,000, while Munich – the third-largest German city – contains 1,245,000 citizens. Hamburg, the second-largest city (population: 1,705,000), found in the northern extremes of Germany, was not chosen as it is not representative of the south-German climate.

[7] Rome, the Italian capital, located about two-thirds down the Italian peninsula, has a population of 2,652,000, while the northern city, Milan, has a population of 1,301,000.

$$GINI = 100* \left(\frac{2* \sum_{i=\text{first household}}^{\text{last household}} \left(HG004_i * HD001_i * VAR_i * \sum_{j=\text{first household}}^{\text{household}} HG004 * HD001 \right) - \sum_{i=\text{first household}}^{\text{last household}} (HG004_i * HD001_i)^2 * VAR_i}{\sum_{i=\text{first household}}^{\text{last household}} HG004_i * HD001_i \sum_{i=\text{first household}}^{\text{last household}} HG004_i * HD001_i * VAR_i} - 1 \right)$$

where:

HD001 = Household size
HG004 = Household cross-sectional weight
VAR = Total net household income divided by equivalised household size (OECD)[8]

Deprivation is calculated using a composite index of multiple indicators of material and social deprivation, as discussed in Chapter 3 and detailed in Healy (2002b). Levels of fuel poverty are taken from a recent pan-European analysis of fuel-poor households in which estimates were calculated using a suite of indicators of housing conditions, affordability of home heating and energy-efficiency levels based on a consensual approach.[9] Data on thermal-efficiency standards of European housing were obtained from Eurostat (1999); Norway and Sweden are included in this analysis as thermal data were not available for Spain or Italy.

Table 9.1 Weather stations employed for climatic data

Country	Weather Station(s)
Austria	Vienna
Belgium	Uccle (near Brussels)
Denmark	Copenhagen
England	London
Finland	Helsinki
France	Paris and Marseilles
Germany	Berlin and Munich
Greece	Athens
Ireland (Republic of)	Dublin
Italy	Rome and Milan
Luxembourg	Luxembourg
Netherlands	Amsterdam
Northern Ireland	Belfast
Portugal	Lisbon
Scotland	Glasgow
Spain	Madrid
Wales	Aberporth (near Cardiff)

---

[8] Thanks to Finbarr Brereton for his help with this equation. Readers should consult Eurostat (1996) for more on the derivation of the Gini coefficient.
[9] Reported in Chapter 3.

Results

Excess Winter Mortality in EU-14

The results show that, between 1988 and 1997, Portugal has the highest seasonal variation in mortality in Europe, with a winter increase of some 28% above the average mortality rate; this amounts to some 8,800 premature winter deaths each year. Ireland also fares particularly poorly with an increase of some 21% (or 2,000 excess winter deaths annually). England, Wales, Northern Ireland and Scotland – both collectively (as the UK) and separately – all share very high seasonality coefficients. The highest level is found in England (19%, 31,000 excess deaths), followed by Wales (17%, 1,800 deaths), Northern Ireland (also 17%, 800 deaths) and Scotland (16%, 3,100 deaths). Overall, the UK exhibits an average seasonality rate of 18% which represents about 37,000 annual excess winter deaths. Spain and Greece also exhibit similarly high rates of relative excess winter mortality (18% respectively), representing 5,700 premature winter deaths in Greece and almost 19,000 deaths in Spain per annum. Furthermore, Italy demonstrates a level of 16% which accounts for some 27,000 excess deaths. Conversely, Finland Germany and the Netherlands appear to suffer far less from excess winter mortality. Confidence intervals of the mean coefficients are also reported in Table 9.2.

Table 9.2 Relative excess winter mortality in EU-14 (1988-97)

	CSVM	95% CI
Austria	0.14	(0.12, 0.16)
Belgium	0.13	(0.09, 0.17)
Denmark	0.12	(0.10, 0.14)
Finland	0.10	(0.07, 0.13)
France	0.13	(0.11, 0.15)
Germany	0.11	(0.09, 0.13)
Greece	0.18	(0.15, 0.21)
Ireland	0.21	(0.18, 0.24)
Italy	0.16	(0.14, 0.18)
Luxembourg	0.12	(0.08, 0.16)
Netherlands	0.11	(0.09, 0.13)
Portugal	0.28	(0.25, 0.31)
Spain	0.21	(0.19, 0.23)
UK	0.18	(0.16, 0.20)
EU-14	0.16	(0.14, 0.18)

Excess Winter Mortality and Climate

Mean winter measurements of environmental temperature, rainfall and humidity are analysed against the results for relative excess winter mortality. The results demonstrate that climatic variables such as mean winter environmental temperature and mean winter precipitation are found to be positively correlated with levels of relative excess winter mortality in Europe, while relative humidity exerts little relationship. A highly significant regression coefficient of 0.27 is found (P<0.001) with regard to environmental temperature, and Fig. 9.1 illustrates the linearity of the relationship. This positive relationship can been termed the 'paradox of excess winter mortality'. The paradox consists of the fact that higher mortality rates are found in less severe, milder winter climates where, all else equal, there should be less potential for cold strain and cold-related mortality. This result indicates that the typical, inverse relationship normally found between cold exposure and rates of (all-year) mortality does not hold for excess winter mortality. Housing standards have been linked as a potential, causative factor behind this paradox (Clinch and Healy, 2000c; Eurowinter Group, 1997). Countries with relatively warm all-year climates tend to have poor domestic thermal efficiency. Because of this, these countries find it hardest to keep their homes warm when winter arrives. This is especially the case in Portugal, Spain and Ireland, where winter temperatures are relatively mild and excess mortality rates in winter are very high. Conversely, countries with severe climates – such as those in Scandinavia – have to maintain high levels of thermal efficiency, as temperatures demand that houses must retain warmth.

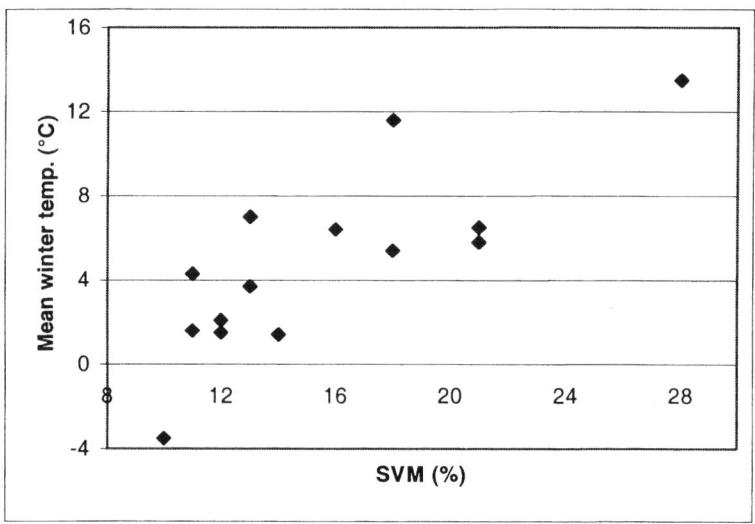

Figure 9.1 Seasonal variation in mortality (SVM) and mean winter temperature in EU-14

Studies on the relative importance of damp (or humidity) on mortality rates are less frequently found, though some associations have been found with impaired health generally (see Chapter 7). This study finds some relationship between the overall level of relative humidity and relative excess winter mortality across Europe; a significant regression coefficient of 0.23 (P=0.02) is reported. The relationship between mean winter rainfall and excess deaths is also found to be significant (a regression coefficient of 0.54, P<0.001).

Table 9.3 Relative excess winter mortality and climatic factors in EU-14

	CSVM	Mean winter temperature (°C)	Mean winter rainfall (mm)	Mean winter relative humidity (%)
Austria	0.14	1.4	14	78
Belgium	0.13	3.7	68	85
Denmark	0.12	2.1	41	86
Finland	0.10	-3.5	39	86
France	0.13	7.0	50	78
Germany	0.11	1.6	49	82
Greece	0.18	11.6	51	71
Ireland	0.21	5.8	65	81
Italy	0.16	6.4	80	78
Luxembourg	0.12	1.5	71	83
Netherlands	0.11	4.3	60	87
Portugal	0.28	13.5	100	77
Spain	0.21	6.5	40	73
UK	0.18	5.4	58	84

Excess Winter Mortality and Macroeconomic Factors

The results of the study show that the state of the macroeconomy is strongly associated with the level of excess winter deaths across Europe (P<0.001), as Figure 9.2 indicates. The negative relationship indicates that more affluent countries with higher per-capita GDP (Luxembourg, Germany, Denmark) exhibit lower seasonal variations in mortality. The converse is also true; the four 'cohesion' countries of the EU (Greece, Ireland, Spain and Portugal) demonstrate the largest mortality variations during winter. This study demonstrates that excess winter mortality does not follow non-seasonal mortality in its relationship with public spending on education. Public per-capita expenditure on both primary and secondary education is found to have an insignificant association with variations in excess winter mortality (P=0.07, 0.06 respectively).

Table 9.4 Relative excess winter mortality and macroeconomic variables in EU-14

	CSVM	PPP-adjusted GDP per capita (Mean, 1988-97 $)	Per-capita expenditure on primary education (% of per-capita GNP)	Per-capita expenditure on secondary education (% of per-capita GNP)
Austria	0.14	20100	20	25
Belgium	0.13	20300	16	23
Denmark	0.12	20400	27	34
Finland	0.10	17600	23	29
France	0.13	18800	13	24
Germany	0.11	20600	-	31
Greece	0.18	12200	11	13
Ireland	0.21	14100	14	22
Italy	0.16	18600	19	-
Luxembourg	0.12	27300	-	24
Netherlands	0.11	18800	-	21
Portugal	0.28	12200	18	20
Spain	0.21	13700	14	-
UK	0.18	17700	-	24

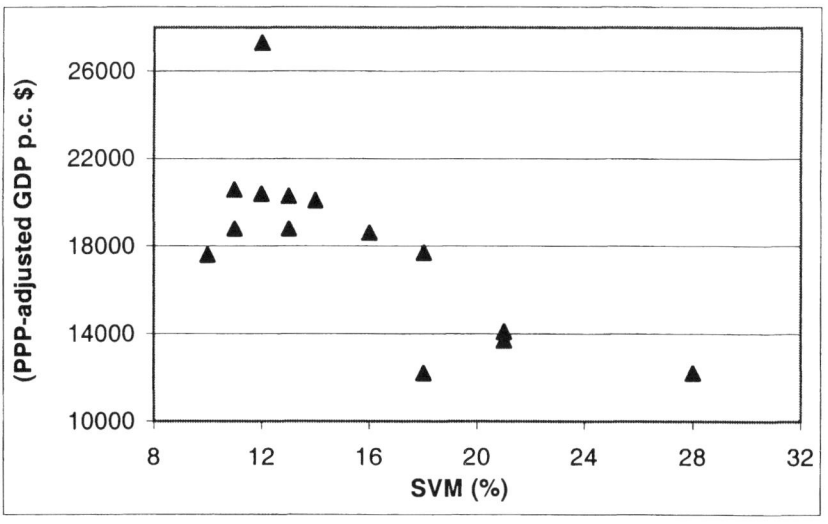

Figure 9.2 Seasonal variation in mortality (SVM) and PPP-adjusted GDP per capita in EU-14

Excess Winter Mortality and Healthcare Provision

This study finds strong associations between various indicators of healthcare provision and seasonal variations in mortality across Europe. Total health expenditure as a percentage of per-capita GNP is found to have a moderate relationship with rates of excess winter mortality. Countries that dedicate relatively high proportions of their national income to healthcare (Germany, France) are found to exhibit lower seasonality in mortality rates than those with relatively low health expenditure (Portugal, Ireland). Disaggregating the data reveals that public-health expenditure is even more strongly associated with seasonal mortality (a regression coefficient of 0.6, P=0.001), while private health expenditure is found to be an insignificant variable in the model (beta coefficient=0.9, P=0.01). Again, countries like Portugal, Ireland and Greece (which all dedicate about 5% or less of per-capita GNP on public-health expenditure) demonstrate the highest variations in excess winter mortality in Europe. Per-capita health expenditure (adjusted for purchasing-power parity) is found to have the strongest association with relative excess winter mortality in Europe, with a regression coefficient of -1.19 (P<0.001). However, the number of hospital beds (per 1,000 population) is found to be insignificantly associated with the coefficient of seasonal variation in mortality (P=0.44), and similarly the number of GPs (per 1,000 population) is not found to be associated with variations in relative excess winter deaths (P=0.67). Figs. 9.3 to 9.6 illustrate the linearity of the significant results.

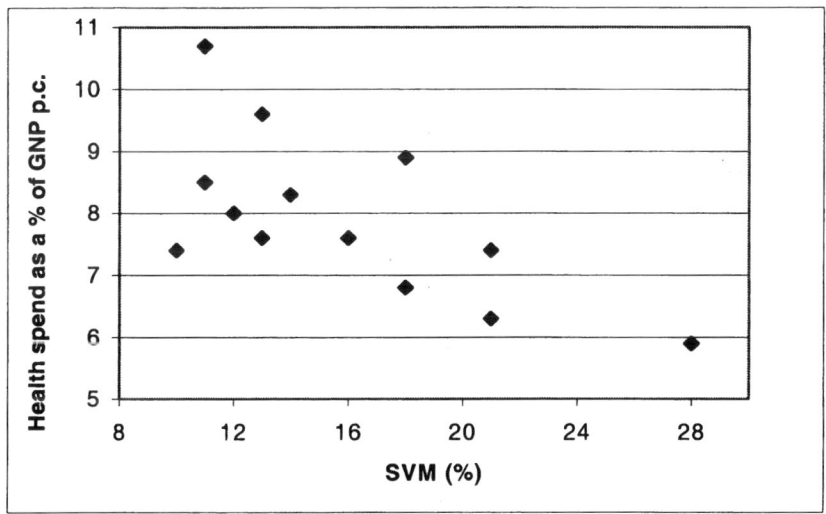

Figure 9.3 Seasonal variation in mortality (SVM) and health expenditure as a % of GDP per-capita in EU-13

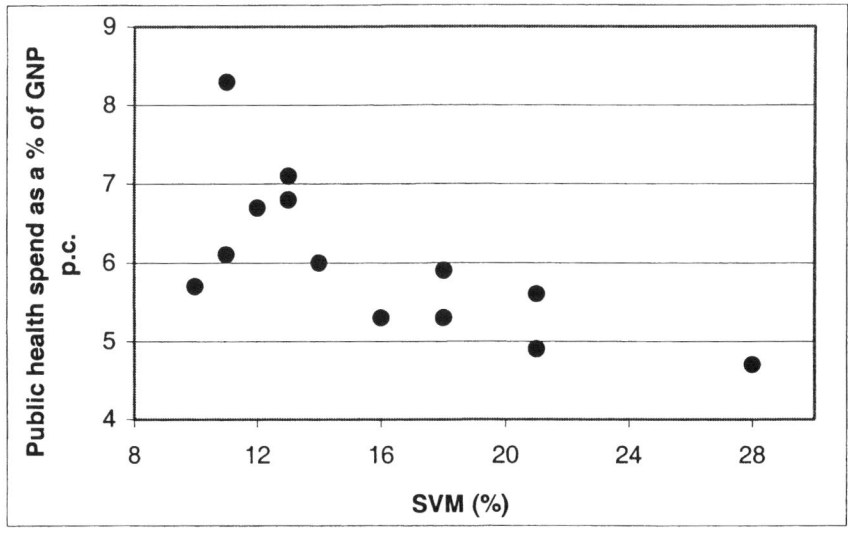

Figure 9.4 Seasonal variation in mortality (SVM) and public health expenditure as a % of GDP per-capita in EU-13

Excess Winter Mortality and Lifestyle Risk Factors

The findings in this section indicate no relationship between lifestyle risk factors and levels of seasonal mortality across Europe. First, smoking rates are found to be insignificantly related to relative excess winter mortality in 13 countries (P=0.34). This implies that, while smoking is found to be strongly associated with mortality and health status using non-seasonal mortality rates, it does not appear to be a significant lifestyle risk factor for seasonal mortality rates. The same test is now carried out for obesity levels – another key risk factor for impaired health status and premature mortality. Again, a similar result is found, with no apparent association between obesity and excess winter mortality (P=0.51). Thus, the hypothesis that the level of excess winter mortality is associated with lifestyle risk factors is rejected.

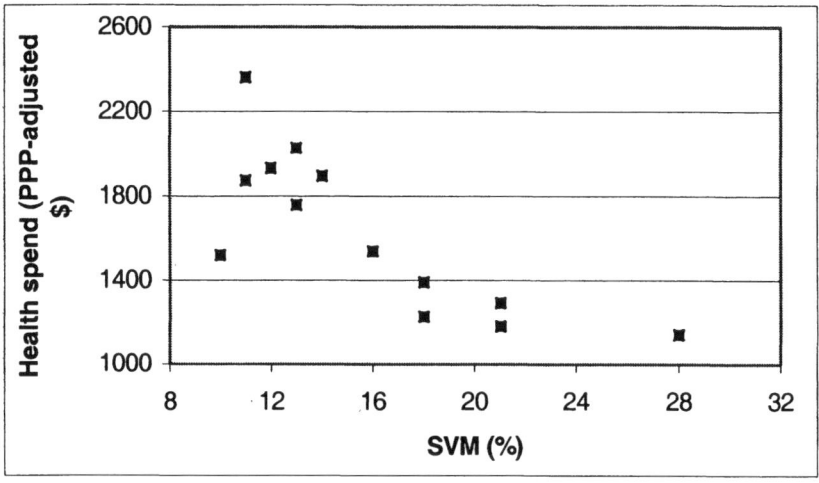

Figure 9.5 Seasonal variation in mortality (SVM) and public health expenditure (PPP-adjusted $) in EU-13

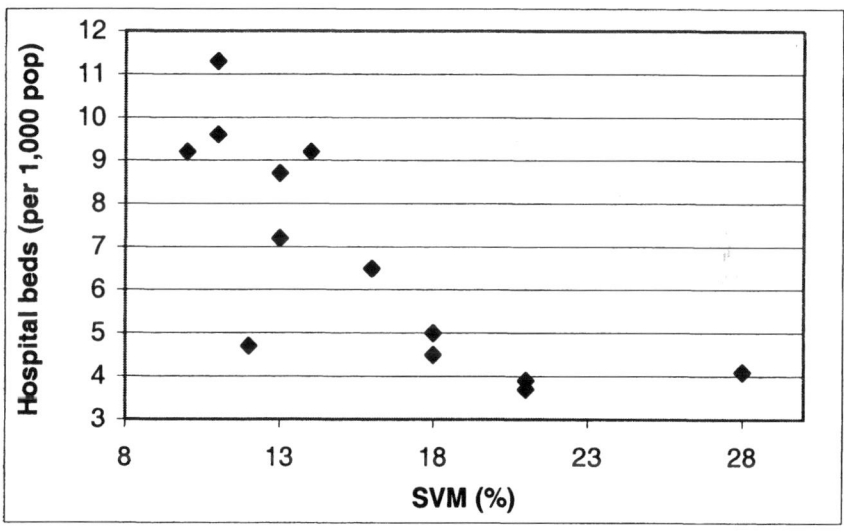

Figure 9.6 Seasonal variation in mortality (SVM) and hospital beds (per 1,000 population) in EU-13

Table 9.5 Relative excess winter mortality and lifestyle risk factors in EU-14

	CSVM	Smoking rate (%)	Obesity rate (%)
Austria	0.14	35	6
Belgium	0.13	25	8
Denmark	0.12	37	6
Finland	0.10	23	9
France	0.13	34	5
Germany	0.11	30	5
Greece	0.18	37	11
Ireland	0.21	29	6
Italy	0.16	32	5
Luxembourg	0.12	-	8
Netherlands	0.11	33	5
Portugal	0.28	27	8
Spain	0.21	37	8
UK	0.18	27	10

Excess Winter Mortality and Socio-Economic Factors

The study now examines four key socio-economic indicators across EU-14. The latest data from the longitudinal European Community Household Panel (1994-97) are employed to examine levels of income poverty, income inequality, multiple deprivation and fuel poverty. The coefficient of seasonal variation in mortality is re-calculated for this section using a comparable time-series, but relative excess winter mortality remains relatively static, although Greek data demonstrate a 4-percentage-point fall in the level of seasonality. Strong cross-country relationships are found between excess mortality and relative income poverty (P=0.01), income inequality using the Gini coefficient (P=0.02), composite, multiple deprivation levels (P=0.05) and composite levels of fuel poverty (P=0.01). Countries with high levels of income poverty and inequality (Greece, Ireland, Portugal) also demonstrate the highest coefficient of seasonal variation in mortality.

Table 9.6 Relative excess winter mortality and health indicators in EU-12

	EWM	Total health spend (% of GNP p.c.)	Public health spend (% of GNP p.c.)	Private health spend (% of GNP p.c.)	Health spend per capita (PPP $)	GPs (per 1,000 pop.)	Hospital beds (per 1,000 pop.)
A	0.14	8.3	6.0	2.2	1896	2.8	9.2
B	0.13	7.6	6.8	0.9	1759	3.4	7.2
DK	0.12	8.0	6.7	1.3	1931	2.9	4.7
FIN	0.10	7.4	5.7	1.8	1520	2.8	9.2
F	0.13	9.6	7.1	2.5	2026	2.9	8.7
D	0.11	10.7	8.3	2.5	2364	3.4	9.6
EL	0.18	8.9	5.3	3.6	1226	3.9	5.0
IRL	0.21	6.3	4.9	1.5	1293	2.1	3.7
I	0.16	7.6	5.3	2.3	1539	5.5	6.5
NL	0.11	8.5	6.1	2.3	1874	2.6	11.3
P	0.28	5.9	4.7	3.2	1142	3.0	4.1
E	0.21	7.4	5.6	1.8	1182	4.2	3.9

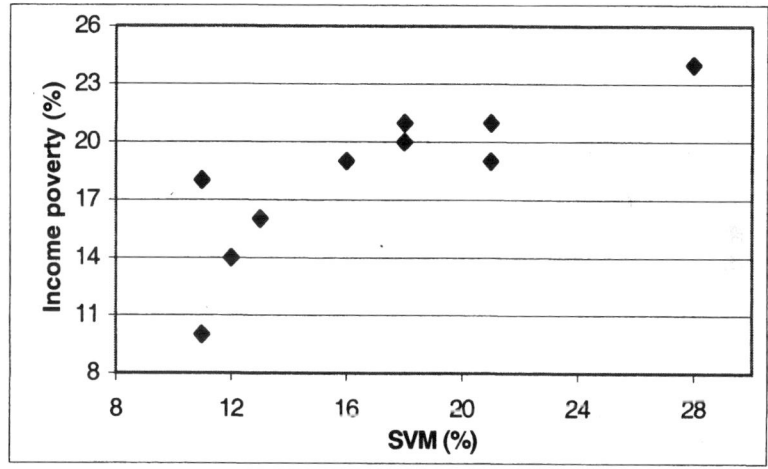

Figure 9.7 Seasonal variation in mortality (SVM) and income poverty in EU-12

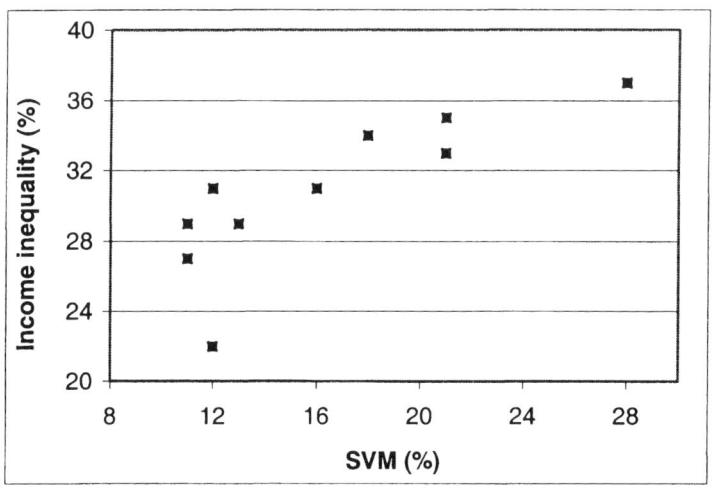

Figure 9.8 Seasonal variation in mortality (SVM) and income inequality in EU-12

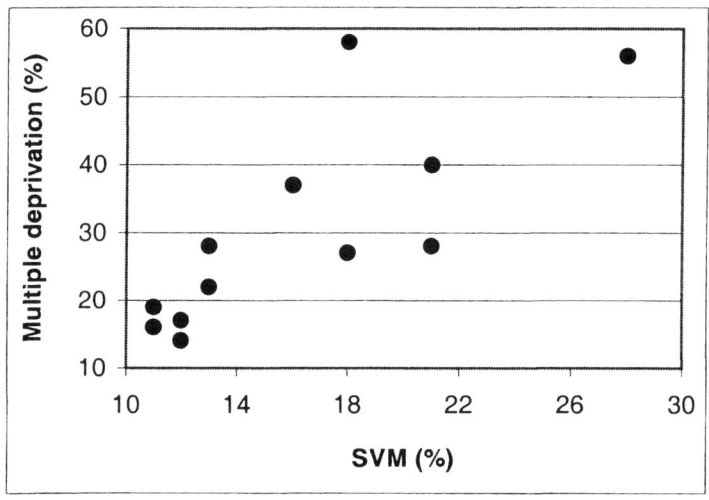

Figure 9.9 Seasonal variation in mortality (SVM) and multiple deprivation in EU-12

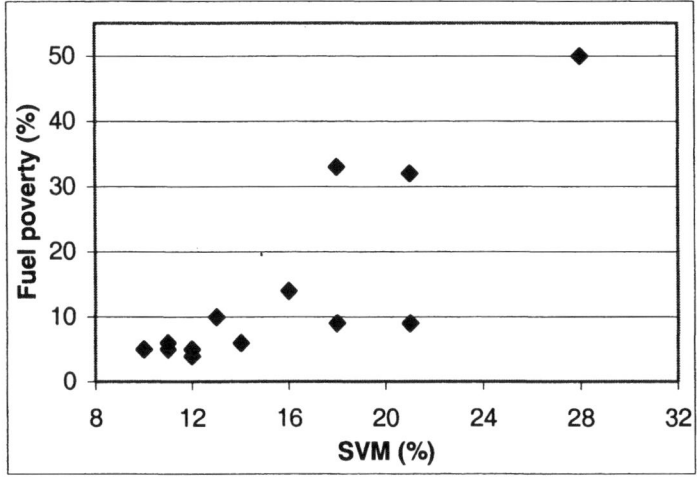

Figure 9.10 Seasonal variation in mortality (SVM) and fuel poverty in EU-14

Figures 9.7 to 9.10 illustrate the degree of linearity found between seasonal mortality and these four socio-economic indicators.

Table 9.7 Relative excess winter mortality and socio-economic variables in EU-14

	CSVM	Income-poverty rate (%)	Income inequality (Gini)	Deprivation rate (%)	Fuel-poverty rate (%)
Austria	0.14	-	-	-	6
Belgium	0.13	17	29	22	10
Denmark	0.12	11	22	17	4
Finland	0.10	-	-	-	5
France	0.13	16	29	28	10
Germany	0.11	18	29	19	5
Greece	0.18	21	34	58	33
Ireland	0.21	21	35	28	9
Italy	0.16	19	31	37	14
Luxembourg	0.12	14	31	14	5
Netherlands	0.11	10	27	16	6
Portugal	0.28	24	37	56	50
Spain	0.21	19	33	40	32
UK	0.18	20	34	27	9

Excess Winter Mortality and Household Thermal Efficiency

If housing standards have been at least part-blamed for levels of excess winter mortality evident in western Europe (Clinch and Healy, 2000c; Eurowinter Group, 1997), then it is useful to analyse data on domestic thermal efficiency to identify if there is an empirical relationship between variations in seasonal mortality and differing housing standards. Table 9.8 demonstrates the results of this exercise. Exemplary levels of thermal efficiency are found in Scandinavian countries. Sweden, Norway and Finland have very high energy-efficiency standards in their homes to combat the relatively severe outdoor environments experienced in these countries.

Table 9.8 Relative excess winter mortality and domestic thermal efficiency in EU-13

	CSVM	Cavity wall insulation (% houses)	Roof insulation (% houses)	Floor insulation (% houses)	Double glazing (% houses)
Austria	0.14	26	37	11	53
Belgium	0.13	42	43	12	62
Denmark	0.12	65	76	63	91
Finland	0.10	100	100	100	100
France	0.13	68	71	24	52
Germany	0.11	24	42	15	88
Greece	0.18	12	16	6	8
Ireland	0.21	42	72	22	33
Netherlands	0.11	47	53	27	78
Norway	0.12	85	77	88	98
Portugal	0.28	6	6	2	3
Sweden	0.12	100	100	100	100
UK	0.18	25	90	4	61

However, available data on southern- and western-European thermal standards illustrate low penetration of energy-saving measures such as cavity-wall insulation, roof insulation, floor insulation and double-glazed windows; this is especially the case in Portugal and Greece, though Ireland and the UK also fare relatively poorly. Strong regression coefficients are found between cross-country variations in seasonal mortality and floor insulation ($R=1.01$, $P=0.03$) and, in particular, double glazing ($R=-0.31$, $P=0.02$); such results may be income-driven. Levels of cavity-wall insulation are found to be related to cross-country variations in excess winter mortality ($P=0.02$). However, the Poisson regression model indicates that roof insulation is not associated with mortality in the model, ($P=0.11$). Note that the degree of linearity between excess winter mortality and the diffusion of certain energy-efficiency measures is not very good, and this is borne out in the regression results, some of which identify insignificant relationships. Figures 9.11 and 9.12 illustrate the linearity of the relationships between variations in seasonal mortality and the penetration of cavity-wall insulation and double-glazing respectively.

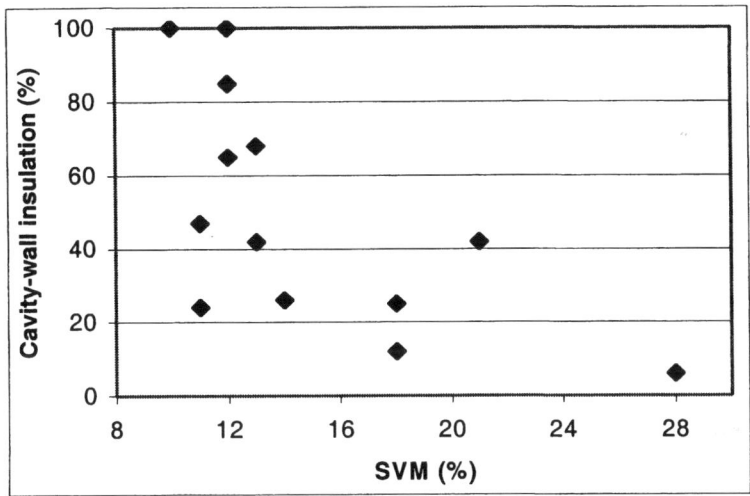

Figure 9.11 Seasonal variation in mortality (SVM) and penetration of cavity-wall insulation in EU-13

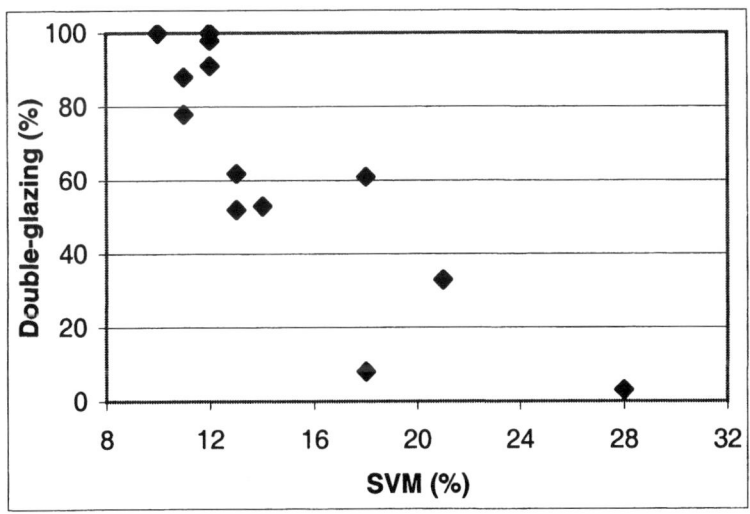

Figure 9.12 Seasonal variation in mortality (SVM) and penetration of double-glazing in EU-13

Discussion

Using data from 1988-97, relative excess winter mortality is found to be highest in southern Europe, Ireland and the UK, where seasonality in mortality of 18-28% is calculated. Scandinavian and other northern-European countries are relatively unaffected by the problem. Such results are startling, especially for southern Europe where virtually no published work exists regarding seasonal variations in mortality. Paradoxically, countries with the mildest winter climates, where mean environmental temperatures remain above 5°C, exhibit the highest variations in seasonal mortality. Other climatic variables, such as mean winter precipitation levels and relative humidity, are not significantly related to cross-country variations in excess winter deaths. However, the strong relationship demonstrated between mean cross-country winter environmental temperatures and levels of relative excess winter mortality indicates that certain populations are far more vulnerable to cold exposure than others. The obvious implication is that both income and housing play strong roles in people's ability to protect themselves from winter temperatures.

Available data on cross-country thermal-efficiency standards in housing indicate that those countries with the poorest housing (Portugal, Greece, Ireland, UK, etc.) demonstrate the highest excess winter mortality. If the ability of a population to protect themselves from cold spells is a key factor in such pronounced seasonality in southern and western Europe, as has been mentioned previously, then it would appear that improving the thermal standards of housing could be an effective preventative intervention in curbing excess deaths. Such a health strategy would also assist in the alleviation of fuel poverty which, this study shows, is also highest in those countries in southern and western Europe with the poorest energy efficiency (Chapters 4 and 5).

Socio-economic indicators of well-being (poverty, income inequality, deprivation and fuel poverty) are also associated with cross-country levels of excess winter mortality. This suggests that levels of excess winter mortality could be reduced through socio-economic progress, as was found in a longitudinal analysis of the Netherlands (Kunst et al., 1990), especially in countries with more unequal income distribution. Macroeconomic data indicate that levels of relative excess winter mortality are associated with per-capita GNP, but not state expenditure on education. Lifestyle risk factors are also not associated with seasonal variations in mortality, unlike all-year mortality rates. However, a number of indicators of healthcare provision are significantly associated with cross-country variations in seasonal mortality. Strong associations are reported for public health expenditure as a proportion of per-capita GNP, purchasing-power adjusted health expenditure per capita and hospital-bed ratios. The latter indicator is especially worrying, as it indicates that potential bed shortages could be resulting in increased seasonal mortality in winter when resources are at their most stretched. Such findings are cause for concern for those countries with relatively low hospital bed ratios and short average durations of stay, most notably Ireland and Spain.

Table 9.9 Results of Poisson regression model: cross-country relationships with relative excess winter mortality (regression coefficients and significance)

	Beta	P-value
Mean winter environmental temperature	0.27	<0.001
Mean winter precipitation	0.23	0.017
Mean winter relative humidity	0.54	<0.001
Purchasing power-adjusted per-capita GDP	1.08	<0.001
Per-capita spend on primary education as a % of per-capita GNP	0.12	0.070
Per-capita spend on secondary education as a % of per-capita GNP	-0.27	0.061
Total health expenditure as a % of per-capita GNP	0.63	0.069
Public health expenditure as a % of per-capita GNP	0.60	0.001
Private health expenditure as a % of per-capita GNP	0.90	0.011
Purchasing-power adjusted per-capita health expenditure	-1.19	<0.001
GPs per 1,000 population	0.59	-0.437
Hospital beds per 1,000 population	-0.23	0.067
Smoking rate	0.40	0.344
Obesity rate	0.30	0.512
Income-poverty rate (60% median equivalised income)	-0.47	0.008
Income inequality (Gini coefficient)	0.97	0.020
Deprivation rate (composite)	0.11	0.048
Fuel-poverty rate (composite)	0.44	0.005
Cavity-wall insulation (% of housing stock equipped)	-2.56	0.022
Roof insulation (% of housing stock equipped)	1.36	0.110
Floor insulation (% of housing stock equipped)	1.01	0.029
Double-glazed windows (% of housing stock equipped)	-0.31	0.024

Conclusions

To sum up, some conclusions can be made:

- High levels of excess winter mortality are confined to southern Europe, Ireland and the UK, all of which endure the least severe winter climates.
- The diffusion of energy-efficiency measures is found to be associated with seasonal variations in mortality.
- Macroeconomic and socio-economic indicators are associated with seasonal variations in mortality, as is the standard of healthcare provision.
- Excess winter deaths could be reduced through improved protection from the cold indoors, increased public spending on healthcare, and enhanced socio-economic circumstances resulting in more equitable income distribution.

In short, this chapter has provided firm evidence that efforts to reduce fuel poverty will produce a positive effect with regard to the incidence of excess winter mortality. However, the regression model indicates that there are several other, equally important, factors that are related to a country's level of excess winter deaths. Therefore, a combination of policies are required to have a desired effect. These are outline on a country-by-country basis in the final chapter.

There are a number of limitations owing mainly to the ecological design of the study. Cross-country analyses are often problematic as standardised datasets are difficult to obtain. A number of assumptions and approximations have to be made when dealing with such large-scale macro datasets and these carry errors. However, great care was undertaken in obtaining the highest quality longitudinal data in an effort to offset potential uncertainties associated with multi-country ecological comparisons.

Although this study has not proven causality, the strong, positive relationship with environmental temperature and the equally strong associations with thermal standards in housing indicate that improving the thermal efficiency of housing in southern and western Europe would play a strong role in reducing the large seasonal variations in mortality found in these countries. The results corroborate the work of multi-centre studies indicating greater impacts on mortality in southern and western Europe than in colder northern regions.[10]

---

[10] Such as that of the Eurowinter Group (1997).

Chapter 10

Policy Implications, Strategies and Recommendations

Introduction

It is now almost 30 years since fuel poverty was first identified formally, a product, in part, of the burden of increased fuel prices resulting from the OPEC oil crises of the 1970s. Since that time, the respective governments of Britain and Ireland have recognised fuel poverty as a *bona fide* social problem. However, many other European countries have yet to acknowledge the existence of fuel poverty, let alone devise and implement fuel-poverty policy. Fuel poverty is arguably the strongest adverse social impact resulting from the inefficient consumption of energy in the domestic sector, and much has been done over the last decade or so both to highlight the plight of those households living in fuel poverty, and to reduce the numbers suffering. Such effort has realised actual results in the form of reductions in the levels of households caught in a fuel-poverty trap in the UK, with 35% less households suffering in 1998 than in 1991 (DEFRA and DTI, 2001). Although declining real energy prices in the 1990s in the UK have played a positive role in this reduction in the level of fuel poverty, some of it may be associated with the Home Energy Efficiency Scheme (HEES) which has provided grants for the retrofitting of energy-saving measures in the home for those on low incomes. However, there has been no similar programme implemented in Ireland, and this study has demonstrated that levels of Irish fuel poverty remain among the highest in northern Europe, while domestic energy-efficiency standards remain poor relative to the rest of northern Europe. Southern Europe appears to have a particularly daunting task in improving the diffusion of energy-efficiency technologies in the domestic sector and curbing fuel poverty.

This chapter presents a round-up of the main policy implications of this study. It then discusses the reasons for market failure and outlines the instruments available to policy-makers to rectify this market failure. Fuel-poverty policy and strategy measures in selected EU countries are then discussed. Finally, the chapter outlines a set of strategies on a country-by-country basis to improve the energy-efficiency of housing and reduce fuel poverty across the EU. Some final conclusions are made at the chapter's close. This chapter is purely an illustrative overview of policy issues surrounding fuel poverty. As such, it does not contain the same depth of analysis as previous chapters, and is not meant to be definitive.

Policy Implications

It is argued in this study that the relatively poor thermal-efficiency standards of Ireland, the UK and southern Europe are serious cause for concern. This is because less well-off households are unable to protect themselves adequately from the cold owing to the energy-efficiency characteristics of their dwelling making home-heating unaffordable. It is clear that the welfare regime and socio-economic characteristics generally play a big role in mitigating fuel poverty. Countries such as Finland, Denmark, Germany and the Netherlands have relatively strong social welfare support and more equitable income distribution, and despite enduring harsh winter climates all report levels of fuel poverty far lower than those found in southern Europe, the UK and Ireland, where income poverty and inequality is relatively high. Ireland demonstrates the strong positive role of economic growth and rising disposable income in reducing the incidence and intensity of fuel poverty. This section outlines some key policy results of the study.

Environmental Policy

A key policy implication of inefficient use of energy in southern and western Europe relates to environmental agreements on stabilisation of greenhouse-gas emissions and acidification precursors (the Kyoto and Gothenburg Protocols respectively). With 1990 as the base year, most countries in Europe are required to reduce their energy-related environmental emissions by 2010, with the notable exceptions being the 'Cohesion' countries which have been allocated an increase over 1990 levels. These policy targets are challenging and require dramatic reductions in business-as-usual levels of energy-related environmental emissions across Europe. It can be seen from this research that there is a significant 'efficiency gap' regarding domestic energy-efficiency in Europe, especially in southern Europe, the UK and Ireland. This is of particular interest for Ireland given that its spectacular economic success over the past decade has made its Kyoto and Gothenburg emissions' targets for 2010 very formidable due to the strong link between economic growth and energy use. The business-as-usual prediction for Ireland is for an overshoot of 25% (10.6 million tonnes) of greenhouse gases and it is clear that Ireland, and other EU Member States, must implement policies over the next decade to curb such environmental emissions; otherwise, a substantial fine will be imposed by the European courts.

Previous research in Ireland indicated that a programme to improve the thermal efficiency of the Irish housing stock to bring it in line with current building regulations would reduce the Kyoto overshoot by 28% and corresponding overshoots in SO_2 and NO_x by 12% and 14% respectively (Clinch and Healy, 2000b). As such, it makes sense on many grounds to improve domestic energy efficiency and reduce fuel poverty.[1]

---

[1] Notwithstanding any 'rebound effect' (i.e. when improved energy efficiency results in reduced demand for energy which then results in reduced energy prices which, in turn, leads

Housing Policy

Housing Conditions The study has shown that serious levels of housing deprivation exist in southern Europe. All four southern countries demonstrate the highest levels of overcrowding in the study (using both objective and self-reported measurements). Furthermore, damp is found to be a serious problem in Portugal, Spain and Greece. These findings have profound public-health repercussions, as damp, in particular, has been shown to be detrimental to human health. Northern-European countries suffer less from poor housing conditions. However, 15% of Irish households are statistically overcrowded and 18% of British households declare a shortage of space. Belgium, France, the UK and Ireland demonstrate the highest levels of poor housing cumulatively and over time using the ten housing indicators. Such results indicate that current housing policies in countries most affected by housing deprivation are not wholly successful in alleviating housing stressors such as overcrowding and damp conditions among vulnerable households. However, the time-series data allow for comparisons over time, and it is clear that most countries in EU-14 are experiencing improved housing conditions during the period 1994-97.

Housing Affordability There are a number of disturbing findings regarding the lack of social equity in relation to housing policy across Europe. First, it was demonstrated that high levels of financial hardship exist in relation to meeting housing costs throughout Europe, and notably in relatively prosperous countries such as Italy, Belgium, France and the UK. High levels of unaffordable housing are also reported in poorer, southern countries like Spain, Greece and Portugal. In addition, state provision of home-heating costs among low-income households appears to be highest in Finland and Denmark. Conversely, tenants in Greece and the UK are far more likely to have to meet the costs of their own fuel bills, and these two countries, unsurprisingly, demonstrate the highest levels of late-payment of utility bills. Longitudinal analysis indicates that housing affordability, unlike general housing conditions, is not improving over time; many Member States, in fact, demonstrate increased financial problems in meeting housing costs over the period 1994-97.

Housing Satisfaction The findings in this study relating to housing satisfaction are also cause for concern. The levels of housing dissatisfaction reported in Greece, Italy and Portugal, in particular, are worrying, for housing is generally regarded as a key indicator for the quality of life of a given population, as well as being fundamental for subsistence. Furthermore, the data indicate that housing dissatisfaction is not demonstrating a marked reduction in EU-14, with some Member States reporting higher incidence over the time series. Policy-makers in these countries should bear in mind that, as housing satisfaction is closely associated with satisfaction with life as a whole, policies to improve housing

---

to an increase in the demand for energy) which would lower the potential reduction in environmental benefits.

conditions will not only assist in achieving environmental-policy targets and reducing fuel poverty and related ill health (much of which is borne by the State), but they should also prove to be politically beneficial measures.

Fuel Poverty and the Fuel Subsidy in Ireland

The study indicated that the fuel allowance is a necessary but insufficient measure in tackling fuel poverty in Ireland. Households claiming the fuel subsidy report an incidence of fuel poverty some three times higher than non-claiming households. However, the fuel allowance appears to impact positively on the severity of experience, significantly reducing the proportion of chronic fuel-poor households in Ireland. As such, the study argues for a continuation of the fuel subsidy.

Reasons for Non-Investment

The study presented the results of the national household survey of Ireland regarding the reasons why energy-saving measures are not adopted by households. The findings indicate a large 'information gap' in the market of domestic energy-efficiency measures, with over a half of households either unaware of the existence or unaware of the benefits of energy-saving measures in the home. In addition, over a third of households identified financial constraints to retrofitting, with only a very small proportion blaming transactions' costs. Thus, financial assistance and improved information and awareness are key policy measures required for the mitigation of fuel poverty in Ireland (discussed in more detail in a subsequent section).

Public Health

Morbidity Fuel-poor households persistently report lower levels of health status (and higher levels of poor or impaired physical and emotional health) than those not so classified. Moreover, a dynamic relationship is reported. Many fuel-poor households are self-aware of their health risks and reduced health status. A major finding of the study relates to housing as a self-perceived causal factor of the levels of poor health status, or more particularly of the levels of chronic diseases. Fuel-poor households are relatively far more likely (three times, in fact) to blame housing conditions as a key cause for their illness than other households. Such results are of strong significance for those responsible for public health in Ireland. As fuel-poor households are generally among the most vulnerable of all low-income social groups, it is likely that most are in receipt of a social-welfare medical card. In this regard, policy-makers should realise that poor thermal conditions in housing and high levels of fuel poverty are associated with higher levels of ill health, the costs of which are being borne predominately by the State. Clinch and Healy (2000d) showed that this excess morbidity associated with domestic energy inefficiency and fuel poverty amounts to an excess exchequer expenditure of €58m in Ireland per annum.

Mortality The study indicated that European countries with poor domestic thermal efficiency and high fuel poverty invariably demonstrate high seasonal variation in mortality and *vice versa*. The majority (over 85%) of this excess mortality occurs in the over-65 age group in Ireland (Clinch and Healy, 2000c). Thus, the public-health implications of fuel poverty are far-reaching, with the potential for premature mortality among the very young and older people.

Quality of Life

The finding that quality of life and life satisfaction as a whole are compromised by the existence of fuel poverty has strong implications for those concerned with social policy and social equity. The data analysis presented in this study indicate that fuel poverty is similar to income poverty and generalised deprivation with regard to its physical and emotional health and well-being effects on those more vulnerable in society.

Why the Market Fails to Deliver Energy Efficiency

The question arises as to why thermal inefficiency and fuel poverty prevail some 30 years after the initial oil shocks lead to the formal recognition of the term. The reasons can be explained by considering the following impediments to effective action.[2]

Policy Constraints in Ireland

The full nature, extent and magnitude of the benefits of domestic energy efficiency in Ireland were a matter for speculation until the Clinch and Healy study (2001). In addition, there is a dearth of comprehensive economic evaluations of energy-efficiency programmes on other European nations, with a large 'grey' literature using mostly benefits-transfers estimates in cost-benefit analyses. The Clinch and Healy evaluation indicated that such programmes are expensive, costing in total (public and private) about €294m (undiscounted) annually over ten years. Heretofore, public finances were such that fiscal rectitude, driven in part by the need to meet the Maastricht Criteria, limited the extent and willingness on the part of the State to embark on substantive investment programmes in energy efficiency. For private households, the recessions of the 1980s resulted in declining disposable income and a subsequent unwillingness to finance new retrofit investment; it is only in the past five to seven years that growth in real household income has been significant (Convery, 1998). This has been the case for a number of countries, not just Ireland. Many European countries like France, Spain, Greece and Portugal are still characterised by relatively high levels of unemployment, income poverty, inequality and material deprivation; this entails that low-income households are unlikely to be in a position to afford improving the thermal efficiency of their

---

[2] Much of this section is based on Clinch and Healy (2000a).

housing in the lack of additional monies. Chapter 6 demonstrated that 32% of energy inefficient households in Ireland cannot afford to undertaken remedial work on their home to improve the level of energy efficiency, while a further 5% have more pressing priorities for funds.

Policy responsibility for energy efficiency in the domestic sector is often spread across several departments and agencies; this has been especially the case heretofore in Ireland. Moreover, Irish energy policy has traditionally focused on supply-side interventions and neglected demand-side options, despite numerous government policy statements to the contrary (Lawlor, 1995 and McSharry, 1993).

Private vs. Social Benefits

Social cost-benefit analysis considers all the benefits to society of a given programme. However, an individual normally only takes account of the direct benefits to him/herself, i.e. the private benefits of energy-efficiency measures. External benefits which are captured by wider society (e.g. reductions in emissions and in morbidity costs to the State) tend not to be considered when a private individual is considering whether to invest in such measures. The payback periods and net benefits of various measures and programmes are adversely affected by the exclusion of non-private benefits.

Moreover, some of the private benefits, such as reductions in the risk of death and sickness from cold exposure and increases in comfort, being non-monetary in nature, are often not considered by the householder when making financial decisions. It is most likely that the householder will consider the cost of the energy-efficiency measures and compare it to the reductions in energy bills they can expect to receive. It is important to note that in Financial Analysis, unlike in Cost-Benefit Analysis, taxes are included because they must be paid by the individual household; however, they are excluded from cost-benefit studies because taxes (and other social transfers) act as distortions in a cost-benefit model.

Market Interest Rate vs. Social Discount Rate

There is no agreement on an appropriate figure for the social rate of discount. In the cost-benefit analysis by Clinch and Healy (2001), a range of discount rates was used and the Irish Government's test discount rate of 5% was employed for the purposes of policy analysis. While this might be considered the appropriate rate for the social cost-benefit analysis, it is less applicable to the private individual. Those who are considering improving the energy-efficiency of their house may not have funds readily available and therefore will be considering taking out a loan from a financial institution. Currently (2002), the nominal rate of interest to be paid on loans is often in excess of 9% in Ireland. At such rates, the net private benefit to an individual becomes negative, i.e. a financial analysis undertaken by the householder would suggest that investment in retrofitting measures would be financially unwise.

Taxation plays a role in the above finding. Retrofitting costs are more expensive to the individual householder than to the state as the individual must pay tax on the costs whereas these are omitted in a social cost-benefit analysis. While energy savings will seem greater to the individual as they will save on paying this tax, labour taxes and Value Added Tax outweigh this benefit such that the investment costs may be several million Euro higher when taxes are included (Clinch and Healy, 2000a).

Socio-Economic Considerations

The least energy efficient households are more likely to be lower income households (Clinch and Healy, 1999a). Such households are much less likely to have available funds and, thus, are most likely to have to resort to a loan. They are less likely to be in the position of accessing credit (particularly at the market rate of interest) and they are more likely to have more pressing alternative uses for any extra funds.[3] They may, additionally, have an aversion to borrowing funds, as has been reported by Salvage (1992). It has also been shown that low-income households tend to have higher discount rates, i.e. they exhibit myopic tendencies whereby they place a greater value on income now as opposed to in the future, partly resulting from the higher degree of uncertainty about the future stemming from their financial instability. Therefore, *ceteris paribus*, such households are unlikely to invest in something that might not pay for itself for over 30 years.

In relation to the policy process, many of those who would benefit most are poor and relatively old and not well represented in the lobbying arena (Clinch and Healy, 1999b).

Information Gap

One of the principal reasons for financially viable energy-conservation measures not being taken up is the lack of knowledge on the part of householders of the opportunities for saving on fuel bills. This information gap is likely to be greater in low-income households where the benefits would be greatest. In addition, an information asymmetry between buyers and sellers of energy-efficiency measures may occur, leading to adverse selection of such technology.[4] Lack of information is seen as a key reason for market failure in the UK according to Williams and Ross (1980) and Carlsmith *et al.* (1990), and this study concurs strongly with British policy analysis of market failure in domestic energy conservation. The results of Chapter 6 showed that one-third of energy inefficient households in Ireland were unaware of the benefits of energy-saving measures, while a further one-fifth, remarkably, were unaware of the existence of such technologies. This means that over a half of energy inefficient households in Ireland (amounting to over 100,000 homes) are suffering from an information deficit with regard to energy efficiency in the home; as such, this appears to be, by far, the strongest impediment to diffusion of energy efficiency in the residential sector. The policy implications of an information deficit on the part of the householder are strong. It is clear that

---

[3] See Weber (1990) for more on this issue.
[4] See Smith (1992).

information and energy-awareness campaigns need to be launched as a matter of course as a means to closing this gap.

In addition, if the housing market worked effectively, the monetary value of the energy-efficiency measures would be reflected in the re-sale value of the house. However, if the public is lacking in knowledge as regards the benefits of the measures, this will not happen. Therefore, if individuals are likely to move house in the meantime, they may not be willing to make an investment with a long payback period.

Transactions' Costs

Another potential 'blockage' in the market for energy-efficiency measures is that of the fixed costs of learning about, and administering, energy-conservation measures. Examples of transactions' costs include the time householders must spend to learn about the various options (information costs, in a sense), locate a suitable installer and oversee the work. Some householders may also be concerned about the appropriate techniques and the quality of the workmanship, as well as the attendant disruption of installing these measures. Such costs are not reflected in Cost-Benefit Analysis and, therefore, the full costs of retrofitting households with energy-conservation measures may be significantly higher to the individual than is suggested by the figures. The amplitude of these transactions' costs may overwhelm the potential pay-off of such an effort, acting as a performance-inhibiting 'wedge' which prevents the implementation of cost-effective energy-conservation measures in the home. These transactions' costs are difficult to measure, but have been seen as potential factors in explaining the slow take-up of financially viable measures in the UK,[5] especially in the domestic sector. The results of this study do not corroborate the hypothesis that transactions' costs act as a major impediment to the diffusion of energy-saving measures in Irish households. In fact, just 3% of energy inefficient households blame such costs as the major reason for not installing these measures.

In addition, the absolute benefits per household are relatively small. Clinch and Healy (2000a) showed that, when the value of all the energy savings of the energy-conservation programme under evaluation in Ireland were added together, they amount to an average of almost £200m per annum, undiscounted over 30 years. However, spread over the number of households, the mean financial gain per household is small at about £163 per annum. In addition, low and (until recently) declining real energy prices, making the energy budget a falling share of total household expenditure, may also act as a barrier; this hypothesis is explored formally in Hassett and Metcalf (1992).

---

[5] This slow diffusion of apparently cost-effective energy-conservation technologies across households has been denoted the 'energy paradox' by Jaffe and Stavins (1994a, 1994b).

Property Rights Failure

Some of the least energy efficient houses in the UK are tenant-occupied (Boardman, 1991 and Brechling and Smith, 1992). The same would appear to be the case for Ireland, as was shown in Chapter 6. Tenants may feel that they are not responsible for undertaking investments in energy efficiency or authorised to do so. Indeed, it is not financially sound for a tenant to invest if they expect to move out in the short to medium term. Likewise, landlords may feel that the benefits to them of such investment may not be recouped if they are unable to raise rents. Also, if investment does take place in a multi-occupancy dwelling, 'free-rider' incentives may exist in relation to the financing of the public good (Smith, 1992).

Summary: Reasons for Non-Take-Up of Energy-Efficiency Measures

There are a number of reasons why energy-conservation measures may not be taken up by the private household: such a household is unlikely to take into account all the benefits to themselves and to wider society of such measures; they may have to borrow funds at an interest rate that would make the investment prohibitive; they may not be aware of such energy-saving measures; the transactions' costs of installing such measures may render the investment unwise. Moreover, the households which would benefit most from the installation of more energy efficient technologies are: least likely to make such a long-term investment; more likely to have to borrow funds (often at a rate of interest higher than the market rate); more likely to have more pressing priorities for extra funds; likely to find it more difficult to obtain such funds; less likely to be aware of energy-efficiency opportunities; less likely to live in their own house.

Policies to close the gap between the positive social benefit of the installation of energy-efficiency measures and the negative private benefit of such measures must therefore endeavour to:

- Close the information gap (among households) and information asymmetry (between buyers and sellers)
- Reduce the opportunity cost of investing funds in energy-conservation measures
- Make such funds more widely available
- Make private benefits reflect more closely the social benefits of such measures
- Reduce the transactions' cost of such investments

Instruments Available to Policy-Makers

There are a number of instruments available to policy-makers to correct for market failure. These are now outlined.

Regulation

Regulation, also known as command-and-control, endeavours to improve the performance of the market via the setting of standards (e.g. building regulations). Non-compliance with a standard results in a penalty, usually in the form of legal action and/or fines. Regulation is likely to be most effective for new housing where minimum standards can be set for insulation. However, it could be mandatory that energy-conservation measures be installed each time a house is sold. It might also be required that information on the energy efficiency of a house (energy rating) be issued whenever a house is sold (see 'Information' below). Landlords could be required to provide minimum heating standards and/or specify the thermal characteristics of the residence to potential renters.

Taxes and Charges

Environmental taxes and charges are economic instruments. These instruments are put in place by a policy-maker to alter market signals to encourage or discourage certain activities or behaviour. A tax on energy generated from fossil fuels may be part of a strategy to reduce emissions of greenhouse gasses. This would provide an incentive to invest in energy-conservation measures. However, energy tends to be price-inelastic and so, when the substitutes for energy generated from fossil fuels are limited, such a tax on its own may not be effective unless combined with other policy instruments.

Tradeable Permits and Offsets

Emissions Trading is also an economic instrument. Rather than being a price instrument (like a tax), it is a quantity-based instrument, whereby compliance with greenhouse emissions or other quotas can be achieved, in part, by purchasing from others whose emissions are below the quota they hold. A price emerges for the exchanges which reflects the scarcity value of the environment. If such a trading system is put in place, it may be possible for emitters who emit a low level of greenhouse gas emissions to sell the carbon reduction to a company that requires emission credits. Such a system will increase the incentive to invest in energy efficiency.

Information

The failure of the market to provide information on the benefits to the householder of energy efficiency can be corrected by improved information provision by the government (see 'Institutional Development' below). As such, information provision can be considered an economic instrument. Provision of information on the benefits of improvement, in the form of an easily read leaflet and a list of installation companies etc., would substantially reduce the information deficit (Convery, 1998). As mentioned above, the inclusion of an energy rating in the specifications of a house on the market could be quite effective as could the provision by landlords of information regarding the thermal characteristics of the residence available to rent.

Subsidies and Tax Relief

Removal of subsidies, if any, on energy-efficiency products would enhance the incentives for energy efficiency. Tax relief (e.g. on the costs of retrofitting) and grants for energy-conservation measures in homes by the government are other potential instruments.

Voluntary Approaches

A voluntary agreement by estate agents that information on the thermal specifications of houses be included in sales literature could have potential. While voluntary agreements by firms to reduce environmental emissions have been shown to work, in the absence of other incentives it may be difficult to get individual households to agree voluntarily to install energy-conservation measures in the absence of other incentives.

Institutional Development

Energy efficiency is usually the concern of a number of government departments; in the case of Ireland, this could amount to at least half a dozen State departments and organisations. However, it is helpful if a focal point is established to co-ordinate policy approaches and to lead the information campaign.

Research and Development

The stimulation of research into the best opportunities for energy efficiency is essential. The construction of cost-benefit analyses and the recommendation of appropriate policy responses in often hampered by a lack of data.

Fuel Poverty and Policy in Europe[6]

This section reviews the various policy measures which have been implemented since fuel poverty was first identified after the oil shocks of the 1970s in a number of countries in Europe and in the USA. It becomes apparent quickly from this

---

[6] Some of this section is taken from NEA (1997).

comparison that policies to address fuel poverty in Europe vary in extent and in effect.

United Kingdom

Measures aimed at improving energy efficiency were first introduced in the UK in 1978 with the Home Insulation Scheme providing grants for up to two-thirds of the cost of the remedial work undertaken for improving thermal efficiency in owner-occupier households. Pensioners were allowed higher allowances and local-authority households were included in the scheme from 1979 onwards. A programme of energy conservation in the social-housing sector was also introduced in 1978 under the Local Authorities Energy Conservation Programme. Initially, this scheme was proposed to be a ten-year programme of retrofitting work to insulate attics and draught-proof doors and windows. Local Authorities were to be given a 'ringfenced' amount of money within their housing-improvement programme which was to be dedicated to energy-efficiency work. However, government cutbacks in 1980 resulted in a ten-fold reduction in the level of remedial work performed by local authorities. Statistics demonstrate that just 66,000 social-sector homes were insulated in 1987, while 630,000 dwellings were improved in 1979. It is argued that the ability of local authorities to improve their housing stocks in the UK over the past two decades has been severely diminished owing to the cut in exchequer funding between 1979 and 1997 which amounts to some 75% in real terms. Local-authority tenants were also able to apply for funding to insulate their attics during the period 1980 to 1990. However, the take-up was poor, with just 6% of grants going to local-authority tenants.

Since 1991 a Home Energy-Efficiency Scheme (HEES) has been in existence in the UK. This is based on the 1978 Home Insulation Scheme and allows for means-tested grants to improve the energy efficiency of the household; those on either income or disability benefit are entitled to apply for subsidies. Funding has risen successively over the years, from £23m in its first year (1991) to £73m dedicated for the 1992 scheme, to £75m in 1998 and almost £150m per annum for the latest 2000-2002 scheme. The mean grant per household was £160 up to the current 1999 scheme. However, this is expected to at least double with the latest scheme. Just 8,000 households availed for the grant in 1991, but this rose to 600,000 households in 1995/96 and 500,000 households in 2000-2002, 300,000 of whom are over-60 years' old. It is thought that, for a typical three-bedroom semi-detached property, the Scheme reduces heating costs by up to £600 per annum, depending on the energy-efficiency conditions pre-retrofit. A grant maximum of £700 is now in place, over twice that in the previous scheme. A special programme, entitled *HEES Plus*, is aimed at the elderly and grants of up to £1,800 are available to such households to improve the heating system as well as insulation levels of the dwelling. Limited winter fuel payments are also made available on a means-tested basis.

Local Energy Advice Centres have been in place for almost ten years in the UK under the aegis of the Department of the Environment and the Energy Saving Trust. Such Centres provide independent, authoritative and free information on the scope for energy-efficiency improvements to the domestic and commercial sectors. There are almost 40 centres currently in operation with an immediate catchment of approximately 15 million people. About 26,000 customers were advised during 1993/94 and it appears from *ex post* research that reaction is generally positive, with two-thirds regarding the Centres as "extremely" or "very" useful, and the same proportion of customers stating that they would use the service again (NEA, 1997). Research has estimated that £57m has been saved in fuel bills as a result of implementing this information-provision service which also helps to reduce the transactions' costs of installing energy-saving measures in the home.

British local authorities have been required since 1990 to provide the Department of the Environment with full details of their work on energy efficiency. The British government's 1990 White Paper on the environment specified that a given proportion of local-authority spending on housing must be dedicated to thermal-efficiency improvements. Some local authorities have gone further and beaten the minimum threshold set by Government.

Utilities have also responded to fuel poverty by developing a range of initiatives aimed at reducing the numbers in fuel debt; payment rescheduling and prepayment cards are two such examples. In this regard, the levels of disconnections have fallen by 87% between 1990 and 1997 (NEA, 1997), and prepayment subscribers account for one-in-seven of all electricity customers in the UK. Voluntary organisations are also very much evident in the UK, with Neighbourhood Energy Action (established 1981) and Heatwise (established in 1983) both active voluntary organisations in England/Wales and Scotland respectively. NEA attracts funding from Government, utilities and other private sources. Between 1981 and 1989, NEA undertook remedial energy-efficiency work in 730,000 homes in England and Wales. It also funds research on fuel poverty. The Scottish counterpart, Heatwise, based in Glasgow, has draught-proofed 110,000 Scottish houses over ten years. It receives mainly European funding, especially through the European Social Fund mechanism, with additional monies from the Glasgow Development Agency and the UK Department of the Environment.

Many of the above British statutory policy measures apply to Northern Ireland as well. However, there are some differences. Grants for private sector housing promote energy efficiency indirectly, as all new-built homes in this sector must comply with strong building regulations. Although a domestic energy-efficiency scheme, similar to the HEES in Britain, is also implemented with similar grant thresholds to the equivalent British scheme, a further £60m subsidy was for low-income homes to deal with relatively high electricity prices in Northern Ireland over the period 1996-99.

Ireland

By contrast, relative to the UK, Irish policy measures have been far less aggressive in tackling the issue, notwithstanding similarly high levels of fuel poverty. The fundamental policy measure in place to combat fuel poverty in Ireland has been based on income subsidisation. Income-supplement allowances have been in place since 1942, initially as an emergency measure to enable those living in urban areas to meet high fuel costs associated with wartime shortages and rationing of coal. Concessionary schemes for specific groups on social welfare were subsequently implemented for various fuels; these included electricity in 1969, LPG (1978) and natural gas (1991). Additionally, a supplementary welfare fuel allowance was introduced in 1977 after the oil crises of the 1970s. Over €240m was spent in 2001 on income supplements for 'Free Schemes' in Ireland, €61m of which (25.4 per cent) was spent on the fuel allowance (Table 10.1). Payments increased dramatically in 2002, with some €80.5m expenditure on the fuel allowance, according to provisional data. This figure represents a 70% increase in payments made 10 years ago. In physical numbers, some 374,000 households availed of this fuel allowance in 2001; the figure is likely to be closer to 400,000 households for 2002.

Table 10.1 Total exchequer expenditure on the 'fuel allowance' in Ireland, 1992-2002

Year	€'000
1992	47,432
1993	47,647
1994	50,666
1995	54,276
1996	56,298
1997	57,070
1998	57,004
1999	55,809
2000	55,487
2001	61,136
2002	80,533

There were two intermittent initiatives regarding capital investment in the domestic sector in Ireland. The first was during the period 1980-82, followed by a scheme in 1985-87. Both of these schemes mainly funded improvements in levels of attic insulation in homes nationwide. The initiatives were well subscribed and successful, with 88,000 households (over 10% of all dwellings in Ireland at that time) benefiting from the grants. As such, attic insulation is the one energy-saving measure with which Ireland performs well relative to its European counterparts, as results in Chapter 5 demonstrated. However, the mid-1980s were economically regressive years for Ireland, and the country underwent a period of severe fiscal rectitude in an attempt to control an ever-increasing national debt. In this regard, both programmes were cut during subsequent spending cuts and never restored. Domestic energy-conservation programmes were proposed in the programmes for government in 1993 and 1995 but very little action followed. The latest Green

Paper on Sustainable Energy and the National Climate Change Strategy gave increased power to the statutory body responsible for energy-efficiency awareness, the Irish Energy Centre, to develop fuel-poverty policy and strategy. Exchequer funding for various measures and programmes comes from within the framework of the National Development Plan (2000-2006) which allocated monies to energy-efficiency programmes in housing. The Green Paper is strong on proposals on improving thermal efficiency in the domestic sector, but far less strong on methods of implementation.

- Energy-efficiency is to be promoted through information and awareness campaigns regarding the benefits of energy efficiency in the home.
- The Government is to "encourage" retrofitting of older, less efficient, housing with improved insulation measures to increase their thermal efficiency.
- Households suffering, or at risk, from fuel poverty are to be "protected".
- Enhanced standards for new buildings in the form of updated Building Regulations will be implemented.
- There will be a new emphasis on building and appliance labelling, thereby assisting energy consumers make more informed decisions.
- Back-up will be provided to train and develop the expertise of builders and contractors in areas such as insulation, heating systems, etc.

The Irish Energy Centre (now Sustainable Energy Ireland) runs a yearly energy-awareness campaign. However, the results presented in Chapter 6 indicate an enormous information gap among fuel-poor households with over 50% of such households (120,000 dwellings) unaware of the benefits or existence of such energy-saving measures in the home. The Centre has also begun to commission research into fuel poverty and the identification of those social groups in Ireland most affected.

The Department of the Environment and Local Government has issued a number of building regulations which increased thermal standards of new-built homes successively. Building regulations from 1991, 1997 and 2001 in particular have raised minimum insulation levels across the new-built stock, but of course these requirements have no effect on the existing 1 million dwellings built prior to the regulations, many of which have little insulation in place. Utilities such as Bord Gáis and the ESB have also implemented measures to tackle fuel debt. Approximately one-in-ten gas and electricity customers in Ireland now use prepayment and rescheduling services in an attempt to avoid fuel poverty. Bord Gáis has implemented a system of energy certification for new houses and the ESB has introduced demand-side measures to reduce demand for electricity at peak times when prices are highest by introducing reduced-rate night-time prices and selling energy-saving measures in its shops nationwide.

Energy Action is the main voluntary actor at play in Ireland with regard to improving energy efficiency. Almost 10,000 households have benefited from their services since its foundation in 1988, most of which were in the Dublin area, although projects have been undertaken in regions as remote as Tory Island. Energy Action have also commissioned the most thorough research on the economics of energy efficiency in the domestic sector. The work, undertaken by the Energy Research Group and Environmental Institute, University College Dublin, indicated huge net benefits to society (in the region of €3 billion) resulting from a large-scale programme of improved domestic energy efficiency (Brophy *et al.*, 1999). Heat and Energy Action Tallaght also addresses fuel poverty in a disadvantaged area of south Dublin by bringing the issues to a national level and organising conferences on fuel poverty from time to time.

Netherlands

The Netherlands has been exemplary in tackling domestic energy inefficiency and related problems of fuel poverty. *Ad hoc* grants for insulation were implemented as long ago as 1974 during the first OPEC oil crisis. A national insulation programme was also developed and initiated in 1979, comprised of mandatory insulation for all new-built homes and large grants for retrofitting existing ones. Subsidies covered a broad range of energy-saving technologies, not just roof insulation, and included double-glazing, cavity-wall insulation, floor insulation, roof insulation and draught-proofing. The programme was altered in 1982 so that grants to owner-occupiers ceased and only tenants were targeted. This was to address the relatively low take-up of the scheme among low-income groups and to reduce the numbers of free-riders among upper income households.[7] In 1987 the programme evolved into a house-improvement scheme of which efficiency measures became one key part. A fixed subsidy (up to a maximum ceiling) was available for insulation measures and double-glazing, or alternatively 25% of remedial work on home improvements could be claimed if energy-efficiency improvements were incorporated in the remedial work.

For over a decade, Dutch utilities have charged an environmental levy on energy bills paid by customers with all proceeds being used to subsidise energy-efficiency improvements in the domestic sector. This levy, which is up to 2% of the total cost of bills, is matched by government subsidies which are then earmarked for investment in CHP, retrofitting insulation and so forth. Exchequer spending on energy conservation was also doubled in 1991 under the National Environmental Energy Plan Plus and the Energy Conservation Memorandum to the equivalent of about €16 per capita. A third of the latest budget is devoted to domestic energy-efficiency measures, with particular emphasis on retrofitting insulation. Energy utilities are required by law to promote efficient energy consumption and are made responsible for conservation programmes. The Dutch government has also created a statutory energy agency, similar to the Irish Energy Centre, called Novem. However, budgets are proportionately far higher than those allotted to the equivalent Irish agency, with Novem receiving large R&D funding

---

[7] Free-riders may be defined as those households who took part in the remedial programme and drew down exchequer funds even though they would have undertaken such remedial work in the absence of such State funding.

which has been used in the domestic sector to monitor various state-of-the-art heating systems and evaluate their efficiency with a view to curbing energy use in this sector by 25% over three years.

Germany

In 1978 a comprehensive energy-saving programme was introduced in Germany which enabled households to opt for grants (for tenants) or tax credits (for owner-occupiers) to retrofit insulation and improved heating systems. The total cost of the scheme was then DM4.4m, with a 50:50 split between tax breaks and subsidy payments. The scheme worked as follows: a 25% grant was made available to cover the range of €4,000-11,800 per home, or a tax allowance was given for ten years which amounted to 10% of the total cost of the investment. Innovative, state-of-the-art options were also funded, such as solar energy. However, most of the subsidy (77%) went towards double-glazing. The scheme has been considered cost-ineffective in more recent reviews (Convery, 1998; NEA, 1997), owing to the low energy benefits arising from the retrofitting of costly measures such as double-glazing. Despite the less-than-robust private energy benefits of this scheme, it has been argued here that the programme would be likely to pass a cost-benefit test if the programme evaluation was widened to capture the non-energy benefits to householders, such as improved health status and comfort (Convery, 1998). Current housing strategies in Germany include an element to improve the affordability of home-heating costs and considerable effort is made to reduce transactions' costs for householders wishing to undertake remedial work. In addition, an informational component is integrated into the policy strategy, with many State-funded energy-efficiency advice centres in operation.

Denmark

In 1975, the Danish government introduced grants and incentives for energy-efficiency programmes. The first scheme (1975-80) provided a grant covering 20-35% of the total cost of the remedial work, depending on employment status. Tax credits were also introduced for households investing in energy-efficiency improvements. The cost of the programme amounted to approximately €700m in 1980 and was deemed less than cost-effective; householders were reported to be using the grant for general home-improvements as opposed to those specific to energy efficiency. Furthermore, households not investing in such measures were found to be suffering from myopia, displaying high discount rates and failing to realise the long-term nature of the benefits of improving the thermal efficiency of their dwelling. In 1980 the programme was modified and re-implemented. It allotted an enormous sum of money (about €1.6 billion) and very substantial per-capita grants to improve thermal efficiency in the domestic sector. The fund was rapidly exhausted owing to the large level of subsidisation available to participating households, but was considered more successful using narrow cost-benefit assessment.

Denmark has continued to implement rigorous and aggressively funded fuel-poverty alleviation programmes over the past decade. The Danish Energy Agency – which is the Danish equivalent to the Irish Energy Centre and Novem in the Netherlands – currently implements four major schemes to improve thermal efficiency and reduce fuel poverty in Denmark. First, a 50% grant is available to those homes wishing to install central heating and connect to a CHP generator. Second, a 50% grant is available to pensioners for energy-saving measures such as insulation, lagging jackets, draught stripping, double-glazing and improved heating controls. Third, a fuel allowance is provided to pensioners on a sliding scale: between 25% and 75% of heating costs are covered depending on how much energy is consumed in the dwelling (i.e. the more energy is consumed the less the subvention). Fourth, an urban-renewal programme has been implemented by the Agency which requires that houses refurbished using subsidies by the Government must also undertake energy-efficiency improvements as part of the subsidised remedial work.

Norway

Norwegian housing, like that in Sweden and Finland, is highly energy efficient. The government in Norway has enforced strict building regulations since the 1960s, well before any oil crises were evident. The retrofitting of older dwellings has been encouraged in the private owner-occupier sector by generous tax breaks. Some commentators have argued that, while the tax breaks given for investment in Norwegian energy-conservation programmes have been very successful, a large number of programme participants were free-riders who would have invested in efficiency improvements even in the absence of the programme (Haugland, 1996). In fact, Haugland estimates that 70% of the participants were free-riders.[8] Full State subsidisation of tenant households took place predominately during the 1970s and 1980s. The effect of Norwegian policy measures in tackling fuel poverty has resulted in a situation where almost 100% of Norwegian households are equipped with double-glazing, 88% have floor insulation, 85% have cavity-wall insulation and 77% have roof insulation.

France

Fuel poverty appears to be an issue which is raised periodically in France, particularly in times of economic stagnation when generalised poverty, inequality and deprivation become more widespread. Non-governmental organisations, such as *Secours Catholique*, have employed the theme 'the right to energy' under their social-service campaign strategies. Limited income supplements have been made available and some administrative options are now underway for consumers of electricity and other utilities so as to curb disconnections. Overall, though, France is similar to many southern-European countries in not having recognised fuel

---

[8] An interesting discussion of the 'free-rider' problem can be found in Franz and Weaver (1994).

poverty as a *bona fide* social problem, despite this research indicating that levels of fuel poverty in France are among the highest in northern Europe (see Chapter 5).

USA

The US government does not appear to formally recognise fuel poverty as a distinct entity separate from generalised income poverty. However, they have encouraged energy-conservation and improved residential energy efficiency using mainly fiscal measures such as tax credits as well as subsidisation of low-income households. Thus, it is not fuel poverty *per se* which has driven the US government over the past two-plus decades to encourage improved thermal efficiency in the domestic sector; rather, it is the wish to keep energy demand stable, thereby retaining a level of autonomy over security of supply. The oil crises of the 1970s also placed an onus on the US government to keep energy costs affordable. If it was not possible to rule out a future energy-supply crisis, then the next best step was to reduce reliance on energy so that the US would have increased power over the affordability of fuel costs.

In 1978 the National Energy Act introduced an explicit policy of energy conservation. A credit of 15% against federal income tax liability was permitted for up to $2,000 of qualifying expenditures on insulation and energy-conserving devices (Quigley, 1991). In addition, a 30% credit was given for investment in renewable resources, later increased to 40% in 1980. Before the introduction of these tax credits, the US government had undertaken a direct programme to increase the insulation levels in dwellings occupied by low-income households.

The Energy Conservation in Existing Buildings Act of 1976 authorised the Department of the Environment to develop and implement a national weatherisation assistance programme to assist in achieving a minimum level of thermal efficiency in the homes of low-income households. Funding was relatively generous with a ceiling of $1,600 per dwelling. Throughout 1984, about $1.4 billion in federal funds had been allocated to low-income households. In the state of California alone, some 66,000 homes were weatherised out of a total number of 1.4 million eligible units. However, some commentators have found that these energy tax credits do not lead to more widespread diffusion of energy-saving technologies, again pointing out the issue of free-riding. Walsh (1989) concludes that the effective discount rate may be too small, transaction' costs in the form of bureaucratic paperwork may be too large, and information gaps may exist.

After the second OPEC oil shock of the late 1970s, the Low-Income Home Energy Assistance Program block grant was established to encourage low-cost weatherisation by low-income households. Federal spending on this programme rose from $279m in 1982 to $398m in 1985. In 1986, the Department of the Environment released almost $2.1 billion to all states to continue the retrofitting programmes. Modest subsidies for solar energy investment through the Solar Energy and Energy Conservation Bank have been witnessed during the later part of the 1980s and early 1990s.

Policy Recommendations

Because of the highly multidisciplinary nature of fuel-poverty research, policy implications fall on many policy-makers doorsteps, and Departmental policies need to be formulated in conjunction with each other to avoid intergovernmental stasis and to achieve economies of scale in an effective fuel-poverty alleviation strategy. It is clear that the fundamental public-health implications of this and other research have to be addressed primarily through appropriate interventions in social welfare and housing policy. However a broad proposal can be made which is based on strong economic grounds.

Domestic Energy-Efficiency Retrofits

Research has shown, time and again, that energy-efficiency programmes in the domestic sector make good economic sense using even the narrowest criteria of assessment. An economic evaluation was recently undertaken of an Irish programme to retrofit all energy inefficient homes with energy-saving measures to bring the thermal standards of the housing stock to the latest (1997) building regulations over a ten-year period (Clinch and Healy, 2001). It was demonstrated that, while the costs of such a retrofitting exercise were large (€1.6 billion), the benefits in terms of reductions in energy bills alone outweighed the costs by 1.7 over the 30-year frame of reference. In addition, the researchers indicated that the external benefits in terms of reduced environmental emissions of CO_2, SO_2, NO_x and PM_{10} were all highly significant in physical and monetary terms, notwithstanding potential rebound effects. Such environmental benefits are of particular importance for Ireland, a country with particularly challenging environmental-policy targets in meeting global environmental Protocols set at Kyoto and Gothenburg on global warming and acidification respectively. The benefits to human health of such programmes are also highly significant, with reductions in morbidity and premature mortality as a result of warmer, more affordable homes. Improvements in thermal comfort were also monetised using a proxy for willingness to pay. Overall, a net social benefit of some €3.1 billion was reported and a benefit-cost ratio of 3:1 is calculated. A remarkable internal rate of return of 33% is also found. Such programmes also have relatively quick payback periods, the costs of this study becoming negative after seven years.

However, not all countries examined in this analysis exhibit similar thermal conditions, nor do all countries demonstrate similar levels of fuel poverty and related mortality effects (see Chapter 11). In addition – and perhaps crucially – socio-economic environments vary dramatically across Europe, with levels of income poverty, inequality and deprivation varying by as much as 40 percentage points from country to country.[9] In this regard, policy measures must be tailored to suit the individual country in question so as to maximise the effect of a given policy response.

---

[9] Readers should consult Chapter 9 (especially Tables 9.2 and 9.6) for more on this.

Austria

In Austria, thermal efficiency was found to be fair, with below-average levels of cavity-wall, floor and roof insulation and poor penetration of double-glazing. Levels of fuel poverty and excess winter mortality were both found to be moderate. It is suggested that the State invests in improved information campaigns in an attempt to encourage the diffusion of energy-saving devices among Austrian households. Tax credits may also be incorporated in such a strategy, although it is important to estimate an appropriate level of tax credit and design a scheme which will minimise the numbers of free-riders. Low-income households may need some support in the form of grants; however, in the absence of comparable socio-economic data on Austrian levels of poverty and inequality, such a strategy requires more close examination.

Belgium

Similar results to Austria were found for Belgium with respect to thermal efficiency, fuel poverty and seasonal mortality, though Belgium demonstrated slightly improved levels of domestic thermal efficiency than Austria. The socio-economic environment in Belgium is characterised by moderate levels of income poverty, inequality and deprivation. In this regard, a programme of improved energy efficiency in Belgian housing should offer means-tested grants to vulnerable low-income households. In addition, it is recommended that State investment in information campaigns be continued so that well-off householders may be alerted to the benefits of installing such measures and encouraged to do so using their private funds.

Denmark

The study found that Denmark demonstrated excellent thermal standards in housing, with over two-thirds of all houses equipped with cavity-wall and floor insulation, three-quarters fitted with roof insulation, and almost all with double-glazing and central-heating systems. Fuel poverty is negligible, although excess winter mortality is moderate. It is suggested that current policy measures, detailed earlier in this chapter, be continued. It is also suggested that more information be provided alerting people to the dangers of cold exposure during winter, especially outdoors. The moderate level of seasonality in mortality in Denmark may well be associated with inadequate protection from the cold outdoors rather than cold strain from within the household.

Finland

Finnish housing is exemplary in terms of thermal efficiency, with all homes fully insulated and double-glazed against the cold. In addition, almost all homes are equipped with central heating. It is not surprising, therefore, that levels of fuel poverty and related seasonal mortality are both very low. Current policy measures should be retained to maintain these high standards and refurbishment programmes should be undertaken from time to time to replace worn-out insulation measures.

France

Although France displays reasonably good levels of thermal efficiency in the domestic sector, with only poor levels of floor insulation, it exhibits a rate of fuel poverty among the highest in northern Europe. The level of relative excess winter deaths is also in the higher end of the range calculated across EU-15. It is suggested in this study that French policy-makers begin investing funds, as a matter of course, in the housing stock to improve domestic energy efficiency and reduce fuel poverty and associated ill health. Because of the relatively high levels of inequality and high levels of deprivation found among French households, it is suggested that substantial subsidisation of low-income households is part of any policy measure to tackle fuel poverty and domestic energy inefficiency in France. Information campaigns may be used to good effect for the more well-off to encourage retrofitting using private funds. If exchequer funding allows, a tax credit may also be used successfully if an appropriate level is assigned.

Germany

Although German housing is not the most energy efficient in Europe, fuel poverty is very low and excess deaths are low to moderate. The relatively strong welfare system in place in Germany, and the attendant moderate levels of income poverty and deprivation, also play a part in the results presented in this study. Thermal standards should be improved, however, even if the main purpose is not to reduce fuel poverty. An inefficient housing stock is consuming greater quantities of fuel than necessary which results in excess emissions of environmental pollutants, such as CO_2, SO_2 and NO_x; all of these pollutants have strong environmental-policy significance, and Germany is legally bound to stabilising emissions of such pollutants under climate-change and acidification agreements (Kyoto and Gothenburg). It is suggested that tax credits be employed in an effort to improve the levels of cavity-wall, floor and roof insulation. Improved information provision may also be effective in such a strategy. Vulnerable households may be targeted for State assistance.

Greece

The results of the study demonstrate that Greece exhibits among the poorest domestic thermal efficiency in Europe, with negligible levels of floor insulation and double-glazing and poor levels of cavity-wall and roof insulation, although about half of Greek households have central heating. One-in-three households in Greece are fuel-poor; this is the second-highest level of fuel poverty in Europe. Associated seasonal mortality is also relatively very high, with an 18% increase in deaths during winter. A full-scale energy-efficiency programme is required to improve the thermal standards of Greek households in an effort to allow households to afford adequate home heating during the winter months. It is obvious that a country like Greece will need to subsidise heavily its contingent of low-income households. This is because income poverty is among the highest in Europe, as is income inequality, while over a half of households exhibit multiple deprivation indicators (Chapter 11). Generous grant schemes must be provided to vulnerable households to assist in undertaking remedial work, otherwise such households will remain caught in a fuel-poverty trap. Informational campaigns may assist upper income households in retrofitting, but it is likely that a tax credit may be needed to encourage such households to carry out remedial work. Such a strategy, however, is unlikely to be implemented unless macroeconomic conditions in Greece change considerably so that exchequer finances would permit such substantial capital outlays by the State.

Ireland

Irish housing is characterised by below-average levels of energy efficiency, with roof insulation being the exception. Fuel poverty is also a problem in Ireland, with Irish households ranked among the most fuel-poor in northern Europe. The level of seasonal mortality in Ireland is the second-highest in EU-15, with a winter mortality rate some 21% above the average rate of mortality. Although some improvements have been witnessed as regards the level of thermal efficiency over the past six years, much of this can be ascribed to the high numbers of new-built homes completed in this period, all of which, theoretically at least, are energy efficient. Just one-in-three households in Ireland are double-glazed and one-in-five have floor insulation, indicating that these measures need to be retrofitted in existing dwellings. The Irish economy has enjoyed economic growth rarely seen in a developed country, with per-capita GDP almost doubling between 1990 and the present (Healy, 2002d). However, it is clear that this wealth has not been distributed particularly equitably, and the Irish have the joint-highest level of income inequality, the second-highest level of income poverty and the highest level of child poverty in the EU (Chapter 11; NESF, 2000). Because of these adverse socio-economic conditions in Ireland, it is clear that high levels of grant aid are required so that low-income and vulnerable households are in a position to improve the energy efficiency of their dwelling and escape fuel poverty.

The fuel allowance should also continue. This research has indicated that, while it may not take people out of the fuel-poverty trap, it alleviates their suffering by reducing the severity of experience of fuel poverty. It is also clear that the Government-funded information strategy, which attempts to persuade households to invest in energy-saving measures using their own private funds, is not working particularly well. Data in this study show that the penetration of energy-efficiency technologies in the home has not increased significantly over the past six years. In addition, the results of the national household survey of Ireland regarding market failure indicate that over 50% of all energy inefficient households remain unaware of the extent of the benefits of energy efficiency in the home; more worryingly, two-fifths of these households are not aware that such measures exist. Quite obviously, there is a large information gap regarding the net benefits of improving energy efficiency among Irish households, and current information strategies appear to be making little, if any, progress in this matter. It is suggested that alternative forms of information provision be explored in an effort to address this information deficit. For campaigns to be successful in changing households' behaviour, a strong message must be conveyed in which the net benefits of energy efficiency in the home are clearly delineated. It may be necessary for Sustainable Energy Ireland to conduct research into the most appropriate forms of information provision and investigate the potential for alternatives.

Italy

Thermal standards in Italian housing remain difficult to assess robustly in the absence of good-quality data. However, it seems reasonable to surmise from other data relating to housing deprivation (presented in Chapter 4) that energy efficiency is not a priority in Italian housing. The relatively high level of fuel poverty in Italy and the 16% variation in winter mortality rates also give weight to this hypothesis. Italy suffers from above-average income inequality and poverty, while deprivation is particularly evident. As such, policy-makers should subsidise households vulnerable to fuel poverty to retrofit their homes to improve the energy efficiency of the dwelling. Information campaigns could be targeted at the more well-off. Tax credits may be a useful incentive, should exchequer finances allow such an economic instrument to be implemented. It is likely that an energy-efficiency programme in the domestic sector in Italy would yield significant reductions in environmental emissions which Italian policy-makers must curb by 2010 under the Kyoto Protocol.

Luxembourg

Thermal efficiency in households in Luxembourg is not among the highest in Europe, however levels of fuel poverty are very low, at less than 5%. The excellent socio-economic environment in Luxembourg, characterised by low income poverty and low levels of multiple deprivation, also assists in maintaining below-average levels of seasonal variations in mortality. However, it is recommended that domestic energy efficiency be improved via information programmes and tax credits to reduce energy consumption and related environmental emissions. Such an outcome would be beneficial in terms of assisting Luxembourg achieve its environmental-policy targets on emissions of greenhouse gases and acidification precursors.

Netherlands

Strong thermal efficiency is found in the Netherlands, particularly with regard to double-glazing and central heating, where four-fifths of all households are so equipped. Despite good thermal standards in Dutch housing, a moderate level of fuel poverty is found in this study in the Netherlands. Seasonal mortality is similar to Germany, with a low-to-moderate rate of 11% found. It is suggested that current policy measures are continued, however it may be useful to examine the potential for introducing a means-tested fuel subsidy (similar to that in Ireland) to reduce the level of fuel poverty further; however, this could result in an increase in greenhouse-gas emissions. Much has been done in the Netherlands to improve the thermal standards of the housing stock, yet a significant number of households report fuel poverty. Moreover, the Netherlands has the lowest level of poverty, the second-lowest level of income inequality and, equally, the second-lowest level of multiple deprivation in EU-14. This suggests that a relatively small number of Dutch households are suffering from a form of deprivation specific to home heating, being unable to afford adequate warmth in the winter. As such, a fuel allowance may achieve a reduction in the fuel-poverty rate in the Netherlands if integrated into the on-going energy-efficiency programme, however there are external costs associated with such policies, as outlined earlier.

Norway

Thermal efficiency is excellent in Norway, though not quite as exemplary as Sweden or Finland, and excess winter deaths are not a substantial problem. This study was unable to estimate fuel poverty, as data on Norway are not collected in the European Community Household Panel, however it is possible to surmise that former and current policies on energy efficiency in housing have been successful in diffusing energy-efficiency measures across the housing stock. The low seasonal mortality found there attests this.

Portugal

Portuguese housing is found to be the most energy inefficient of all housing in the EU. While the country endures the least severe winter of the 15 countries analysed in this study, the negligible insulation levels and poor heating systems in place in Portugal mean that half of Portuguese households are unable to afford adequate warmth in the home during cold winter spells. Furthermore, Portugal suffers from the highest income poverty in EU-15 (with 24% of households below 60% of median equivalised income), the highest income inequality (37 on the Gini-scale), and the second-highest level of generalised deprivation (56% of households reporting multiple deprivation indicators). It is perhaps unsurprising, but nonetheless alarming, to note that the level of related seasonal mortality in Portugal, at 28%, is the highest in the EU by far. Portugal, like Greece, is a relatively poor country, and improving the thermal efficiency of housing does not appear to be a key priority for policy-makers. Current strategies are based on information campaigns encouraging households to invest using their own funds. It is clear that such a strategy is not going to achieve substantial improvements in the diffusion of energy-saving measures in a country where income poverty is rife and fuel poverty affects half the population. Low-income households in Portugal must be heavily subvented to retrofit their homes with insulation and central-heating upgrades. The more well-off should be encouraged through tax breaks and improved information campaigns, so that energy consumption and related emissions are curbed to assist in achieving Kyoto and Gothenburg policy targets.

Spain

Data on Spanish energy-efficiency levels is very difficult to obtain, and Eurostat currently do not have these important data. Available statistics indicate that about two-fifths of Spanish households are equipped with central heating which is better than Portugal but worse than Greece. It could be surmised that Spanish households are insulated to levels between those found in Portugal and Greece. This would appear to be a good conjecture as the level of fuel poverty calculated in Chapter 5 shows that about one-in-three households are suffering in Spain (similar to that found in Greece). Once again, Spain (like Greece and Portugal) is a country characterised by relatively high levels of income poverty, income inequality and generalised deprivation. There is little in the way of fuel-poverty policy in place in Spain and it is recommended that government provide grants to vulnerable, low-income households to enable them to improve the energy efficiency of their dwelling so that they may be able to afford adequate warmth. Again, a tax concession could be introduced to encourage owner-occupiers and upper income households to retrofit. Improved information on the benefits of energy-saving measures would also be beneficial.

Sweden

Swedish households are fully equipped with all insulation measures examined in this study, and all homes are fitted with central heating. Comparable estimates of fuel poverty and other social indicators are unavailable, as Sweden has only just joined the European Community Household Panel. However, seasonal variations in mortality are among the lowest found in Europe, despite having among the coldest winters, and Sweden is among the richest nations in the world using macroeconomic measures such as per-capita GDP and among the least afflicted by income poverty. In addition, the strong social welfare support mechanism entails that income distribution is relatively flat. It is argued that the strict building regulations enforced in the past have ensured that housing standards are high with regard to thermal efficiency in Sweden.

UK

The UK has taken the problem of fuel poverty seriously over the past decade. A Home Energy Efficiency Scheme has been in place since 1991, with several successive modifications providing for increased funding of energy-efficiency improvements in British households. The strategy appears to have been something of a *tour de force*, reducing fuel poverty by one-third between 1991 and 1998. It is likely, however, that some of this progress may be related to the declining real energy prices witnessed over the past decade in the UK. Notwithstanding this apparent success, thermal standards appear to be lagging in British households and are comparable, overall, to those found in Ireland. Although double-glazing had reached a 61% penetration by 1996, floor and cavity-wall insulation remains paltry in the UK. The level of fuel poverty is among the highest in northern Europe, along with Ireland, France and Belgium. Seasonal mortality is also among the highest in the UK relative to northern European levels, with an 18% increase in mortality rates during the winter period. Income poverty, inequality and deprivation are all above the EU-average in the UK. It is suggested that the current HEES policy measure be continued but its funding should be increased. Presently, a target of 250,000 households per annum is in place in Britain, which means that it would take 20 years to retrofit all 5-million households currently in fuel poverty; this assumes full take-up of the scheme which is highly unlikely. Improved information campaigns and tax-concession schemes could also be integrated into the UK's fuel-poverty strategy to encourage the more affluent to retrofit using their own funds.

A summary of the policy recommendations, on a country-by-country basis, is provided in Table 10.2.[10]

Conclusion

Thermal-efficiency standards vary from country to country because of the differing housing policies implemented. Countries with more severe winter climates (especially those in Scandinavia) have, for many years now, invested heavily in energy-efficiency retrofitting programmes in existing buildings while implementing strong building regulations for new-built houses. Such countries are also characterised by strong welfare systems with relatively equitable income distributions and low levels of poverty and deprivation. The problem of fuel poverty, therefore, is not an issue of any great concern in these countries. However, housing policies in countries in western and southern Europe have traditionally not placed great emphasis on thermal efficiency. Building regulations have been less stringent, and investment programmes in thermal efficiency in the domestic sector have generally been *ad hoc* in design. To make matters worse, social welfare systems are less supportive to those vulnerable in society, particularly in southern-European countries. When households in these countries experience a cold winter spell, they are far less well protected from the cold indoors and far less able to afford adequate warmth. These factors have resulted in serious problems of fuel poverty in southern and western Europe.

This chapter has outlined the responses of some European countries and the USA in their attempts to improve the energy-efficiency standards of housing over the past three decades. Most policy responses are based on fiscal measures aimed at owner-occupiers, such as tax breaks, as well as grant aid to more vulnerable households. Information and advice is also a common component of the policy measure. It is recommended that each country adopts policies to improve the diffusion of energy-saving measures to suit their own macroeconomic and socio-

---

[10] The qualitative summary of findings in Table 10.2 is based on the quantitative findings of the previous chapters.

economic conditions. It is argued that southern-European nations like Greece, Spain and Portugal need to adopt the most radical policy shifts in tackling energy inefficiency and fuel poverty. Full-scale retrofitting programmes are recommended with heavy subsidisation of low-income households in an attempt to reduce the remarkable levels of housing deprivation, fuel poverty and related adverse health outcomes. The current HEES policy measure in the UK would appear to be somewhat successful in tackling fuel poverty, with one-third fewer households suffering in 1998 than in 1991. However, it is calculated in this study that, even if the scheme works at 100% efficiency and is fully subscribed, it will take over 20 years for all fuel-poor households to be lifted out of the fuel-poverty net. The scheme, therefore, requires more funding.

With regard to the situation in Ireland, it is clear that Irish policy-makers need to decide upon a coherent and comprehensive strategy to deal with domestic energy inefficiency and fuel poverty. Previous limited *ad hoc* energy-efficiency programmes were highly successful in improving the diffusion of roof insulation in Irish households, however a more extensive programme is now required to deal with the unsatisfactory level of fuel poverty and below-par energy-efficiency standards in Irish housing. While there was some excuse for the lack of government intervention in this area when macroeconomic conditions were less favourable in Ireland, there is very little excuse now in light of superlative economic growth over the past decade. The gap between rich and poor has widened in this time and it is apparent that government strategy in the area of fuel poverty should be heavily supportive of low-income households. It is heartening, however, to note that Sustainable Energy Ireland (formerly Irish Energy Centre) is about to embark on a programme of retrofit subvention to low-income households; such an initiative is a giant step in the right direction, albeit notwithstanding the limited resources allocated.[11] The fuel allowance should also continue, as the study found that it appears to reduce the severity of experience of fuel poverty among more vulnerable households. In addition, the considerable information gap exhibited among energy inefficient and fuel-poor households (reported in the national household survey) indicates that a far more aggressive information and awareness campaign needs to be executed with more emphasis on the range of benefits of installing energy-saving measures in the home, not just savings in fuel bills.

There has been little discussion in this study on energy taxation. Some comments are now warranted. So-called 'green' / carbon / energy taxes could be

---

[11] Sustainable Energy Ireland recently (in 2002) launched a Low Income Housing Programme Strategy through an extended funding programme. The Extended Funding Programme represents a step in the right direction in that it will be the first major energy-efficiency programme specifically targeted at addressing fuel poverty in low-income households. A managing agent for the Extended Funding Programme has been appointed. The measures to be carried out will initially include cavity wall and attic insulation, draught proofing, hot water cylinder jacket, low energy lamps and energy advice.

highly regressive in the domestic sector, further burdening those on low incomes trying to make ends meet and afford adequate home heating for two reasons:

- Low-income, fuel-poor households spend disproportionately high sums on home heating relative to their household income;
- Such households are more likely to use carbon-intensive fuels to heat their homes.

Carbon taxation is currently being rolled out across the EU, and Ireland is likely to introduce such fiscal measures soon, possibly in its 2005 Budget. Energy taxation makes economic sense and assists policymakers achieving stringent targets on greenhouse gases and promotes a more sustainable use of energy across all sectors of the economy. Unfortunately, as has just been noted, carbon taxes in the domestic sector are highly regressive taxes, and such taxes will most probably increase the levels of fuel poverty in Ireland and across the EU. On these grounds this study recommends that appropriate safeguards must be put in place to cushion the effects of carbon taxation policies on low-income households. It has been suggested by this author [in Healy (2003d)] that the revenue from the implementation of a carbon tax should be recycled for current and capital investment to assist low income households from the inflationary price effects of a carbon tax. Doubling the fuel allowance from its current (2004) levels would cost €60 which amounts to 12% of the carbon tax revenue forecasted using a conservative low-tax option of €7.50/tonne. In addition to current expenditure, it is suggested that capital investment in the housing stock continue (through the SEI low-income housing programme) and be increased using recycled revenues from the carbon tax. The current funding envelope of €7.62m allows for improvements to be undertaken in 18,000 houses (an average investment cost of €423 per home). It is clear from the analysis in this book that up to 226,000 homes may require such an intervention, so there is a considerable gap between the target and the actual incidence of fuel-poor households in need of capital subvention. In conclusion, the carbon tax clearly presents challenges but also real opportunities to policymakers to improve the position of low-income homes affording adequate household warmth.

It is clear from this study that government and market failures are responsible for the high levels of housing deprivation, fuel poverty and related adverse health effects found in southern and western Europe. Failure of governments to rectify the *status quo* will entail that vulnerable households will remain living in cold, uncomfortable housing conditions, exposing themselves and their children to a range of adverse health outcomes. Moreover, the risks to older householders of such housing deprivation have been demonstrated to be potentially fatal.

Table 10.2 Summary of key findings and recommended policy strategies

	Thermal efficiency	Fuel poverty	Excess mortality	Socio-econ environment	Suggested policy interventions
A	Fair	Moderate	Moderate	-	Improve cavity-wall, floor and roof insulation and double-glazing via information and tax breaks.
B	Fair	Moderate	Moderate	Fair	Ditto plus grants to low-income households.
DK	Excellent	Low	Moderate	Excellent	Continue current policies. Provide more information on dangers of cold.
FIN	Excellent	Low	Low	-	Continue current policies.
F	Good	Moderate	Moderate	Fair	Improve double-glazing and floor insulation. Grants to low-income households. Awareness and tax credits.
D	Fair	Low	Low / Moderate	Good	Improve cavity-wall, floor and roof insulation. Info and tax credits. Grants to low-income homes.
EL	Poor	High	High	Poor	Full-scale retrofit programme required. Grants to low-income homes. Information and tax credits.
IRL	Fair	Moderate	High	Fair	Improve double-glazing and cavity-wall and floor insulation. Grants to low-income homes. Information.
I	-	High	High	Fair	Improved energy awareness. Grants to low-income households. Improve information and offer tax credits.
L	Fair	Low	Moderate	Excellent	Improve information and offer tax credits.
NL	Good	Moderate	Low/ Moderate	Excellent	Fuel subsidy to low-income households and improved information on dangers of cold exposure.
N	Excellent	-	Low	-	Continue current policies.
P	Poor	High	High	Poor	Full-scale retrofit programme required. Grants to low-income households. Improve awareness and tax credits.
E	-	High	High	Poor	Ditto.
S	Excellent	-	Low	-	Continue current policies.
UK	Fair	Moderate	High	Fair	Grants to low-income homes for cavity-wall & floor insulation. Information and tax credits.

Chapter 11

Summary and Conclusions

Introduction

The purpose of this work has been to analyse the relationship between housing deprivation, domestic energy inefficiency, fuel poverty and related health impacts. The comparative analysis was developed using large cross-country datasets, and a variety of social indicators pertaining to housing and other variables were examined. The longitudinal data allowed for comparisons over time. This chapter summarises the findings of the eight chapters which presented the empirical findings of the study. A summary of the key results of each chapter is provided sequentially.

Housing Conditions in Europe

The pan-European analysis of housing deprivation, energy-efficiency levels, affordability indicators and satisfaction with housing indicated that there are a number of serious causes for concern. In Southern-European countries, especially Portugal, Spain and Greece, there are substantial problems with leaking roofs. High proportions of households in both Portugal and Greece do not possess hot-water facilities; the vast majority of Greek households lack this basic household amenity. To compound these results further, southern-European households are found to suffer from very high levels of overcrowding, using both a technical (objective) measurement and a self-reported (subjective) measurement. Mediterranean countries are also found to have the least energy efficient housing in this group of 15 European nations, with Portugal and Greece displaying negligible levels of home insulation and double-glazed windows.

Northern-European countries suffer less from poor housing conditions. However, there are some notable exceptions. Irish households are statistically overcrowded and, despite relatively good levels of roof insulation, are generally thermally inefficient, demonstrating among the lowest levels of cavity-wall insulation, floor insulation and the lowest level of double-glazing in northern Europe. British households are similarly under-insulated. Austrian households are not highly energy efficient either, and Germany appears to have below-average insulation levels which imposes significant external costs in terms of excess greenhouse-gas emissions. However, these countries are among the richest nations in Europe (using GDP measures) and the least afflicted by poverty and income inequality, hence such 'efficiency gaps' are not as pernicious in these countries as in the UK and Ireland where poverty and inequality are far greater.[1] Moreover, colder northern countries need more protection from the elements than southern countries which experience far less severe winter climates. As such, the relatively poor thermal efficiency standards of Ireland and the UK are, to some extent, of more concern than the very poor thermal standards of southern countries because the interaction between outdoor and indoor environments may be potentially less pernicious in Mediterranean nations.

A particularly disturbing finding is the level of financial hardship which exists in meeting housing repayments across Europe, and notably in relatively rich countries such as Italy, Belgium, France and the UK. High levels are also reported in poorer, southern countries like Spain, Greece and Portugal. In addition, State provision of home-heating costs appears to be highest in Finland, Denmark, Spain, Portugal and Germany. Conversely, tenants in Greece, the UK and Ireland are far more likely to have to meet the costs of their own fuel bills.

Finally, the highest levels of housing dissatisfaction are experienced, perhaps unsurprisingly, in southern-European countries. Greece, Italy, Portugal and Spain all report strong incidences of dissatisfaction with their housing conditions. In northern Europe the UK, Germany and Ireland are found to be the most unhappy with their housing, though the respective incidences of dissatisfied households are considerably below those reported in southern countries. Such findings are cause for concern for those responsible for developing and implementing policy pertaining to housing and welfare, as large segments of households are reporting unhappiness with something that is regarded as fundamental for subsistence and a key factor in life satisfaction as a whole.

Fuel Poverty in Europe

Chapter 3 presented the results of a cross-country analysis of fuel poverty using comparable data; such comparative empirical analysis is the first of its kind. A new methodology for measuring fuel poverty was proposed which attempts to identify the fuel-poor using a number of social indicators of deprivation and by utilising a composite measurement. It is argued that the proposed new methodology for measuring fuel poverty is superior to the problematic and unscientific Expenditure approach for measuring fuel poverty (households that would need to spend more than 10% of income on home heating), especially in cross-country comparisons where such a measurement may be rendered meaningless. These results compare favourably with those reported by Whyley and Callender (1997) in their provisional analysis of fuel poverty using four social indicators.[2]

Using six objective and subjective indicators in a weighted composite index, fuel poverty is calculated to be highest in southern Europe. Portugal, Greece, Spain and Italy demonstrate the highest levels of fuel poverty, regardless of sensitivity analysis; remarkably, half of households in Portugal report fuel

---

[1] As was demonstrated in Chapter 9.
[2] However, a composite index was not constructed by Whyley and Callender.

poverty. In northern Europe rates of fuel poverty are relatively lower; however, France, Belgium, the UK and Ireland exhibit relatively high incidences (9-10%). Using an Expenditure approach, it has been calculated that 16.4% of British households were fuel-poor in 1998 (or 22.3% if housing costs are deducted) (DEFRA and DTI, 2001), and data from the most recent (1995) Irish Household Budget Survey can be employed to derive a corresponding statistic of 20.7% for Ireland (or 25.0% if housing benefits are excluded from household income). Such results are somewhat higher than the findings based on social indicators, and it is argued that the Expenditure approach may over-estimate *actual* levels of fuel poverty, because the arbitrary 10% income threshold may be too low. As was stated, there appears to be no underlying scientific rationale behind setting the fuel-poverty line at 10% of net income; indeed, the only rationale appears to be based on simplicity and ease of calculation. As the academic literature on fuel poverty is undesirably thin, there has been little debate about the appropriate income threshold. The methodology developed in this study overrides the problems associated with the traditional definition of fuel poverty by moving to a composite measurement based on indicators of fuel poverty.

There are a number of implications of this methodology. First, the estimates of fuel poverty in the UK and Ireland derived under the consensual approach are considerably lower than those estimated using a more traditional approach. This implies that the incidence of fuel poverty may indeed be lower in actuality than that which has been calculated formerly. This positive implication does not mean necessarily that government response to the problem can now be 'scaled down' or stopped. It does entail, however, that a new income threshold should be considered by those wishing to calculate fuel poverty using the traditional approach based on fuel spend in the home. Second, the results indicate serious levels of fuel poverty in southern Europe. This is startling when it is considered that fuel poverty is not even recognised in most European States outside the UK and Ireland.

Lone-parent households are shown to have the highest incidence of fuel poverty across Europe, especially in southern countries where as many as three-quarters are fuel-poor. In northern Europe the highest incidence is found in Ireland and the UK where about one-fifth of lone-parent households are suffering fuel poverty; such results are very high relative to most northern countries. Lone-pensioner households are also identifiable as a risk group, especially in southern Europe where, remarkably, up to 89% of such households are declaring fuel poverty. In northern countries Ireland and the UK, again, display the highest levels in this social group. Households living in apartment complexes also exhibit high incidences of fuel poverty, especially in Ireland, the UK and southern countries. The unemployed are a key risk group, with 81% affected in Portugal. In northern Europe the unemployed in Ireland, France and the UK suffer highest, with 23.3%, 21.3% and 17.7% affected respectively. Tenants demonstrate a far greater risk of fuel poverty than owner-occupiers across Europe, as do those with low levels of educational attainment. Finally, the divorced, separated and widowed display persistently high incidences of fuel poverty in Europe. All of these results are highly significant in the multivariate Probit regression model employed in this analysis.

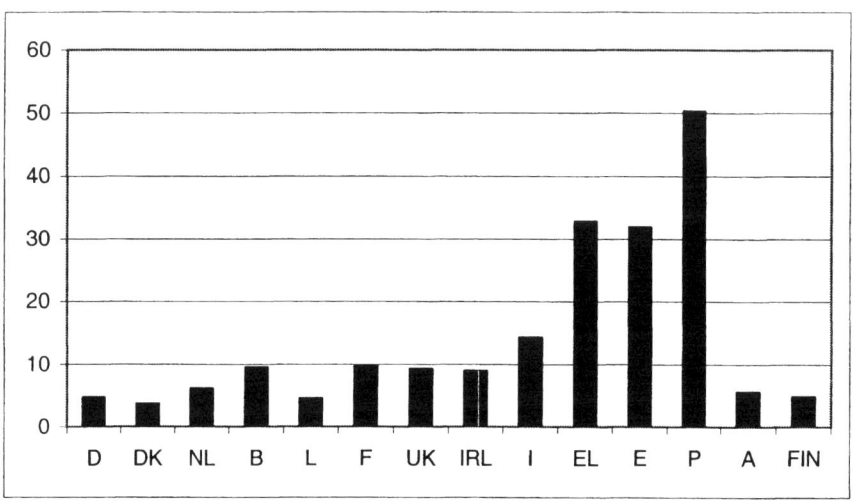

Figure 11.1 Composite fuel poverty in EU-14 (% of households, 1994-97)

Severity of Fuel Poverty in Ireland

This chapter employed data from a national household survey of Ireland in 2001 to assess, for the first time, the severity of Irish fuel poverty and to identify the most vulnerable social groups. It was demonstrated that, while the penetration of lagging jackets, double glazing and central heating have improved substantially over the past five years, much of this improvement is owing to the increased level of new house completions over the past 6 years. Nonetheless, much of the Irish housing stock remains considerably under-protected from the outdoor environment, leaving the vulnerable in society open to fuel poverty and increased risk of ill health. A national estimate of fuel poverty of 17.4% (or 226,000 households) is produced for Ireland using the national household survey. This estimate compares more similarly with the estimate of 20.7% produced from the 1995 Household Budget Survey using the standard Boardman definition of fuel poverty.[3] This research also identified sufferers of fuel poverty by severity. It is estimated that 27% of fuel-poor households (4.7% of the total housing stock) are suffering from chronic fuel poverty, where householders are caught in a persistent fuel-poverty trap, constantly unable to adequately heat the home. It is calculated that 12.7% of all households

---

[3] Where the fuel-poor are defined as those households spending in excess of 10% of disposable income on domestic energy requirements.

suffer from intermittent levels of fuel poverty, where householders are occasionally unable to heat the home adequately.

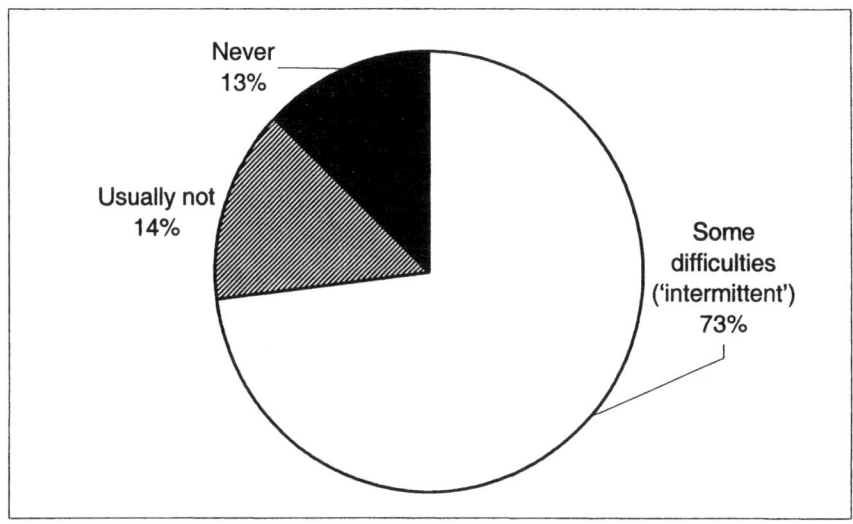

Figure 11.2 Ability to heat the home adequately by severity (% of households, Ireland, 2001)

The apparent discrepancy between the estimate of Irish fuel poverty produced using the national household survey (17.4%) and the European Community Household Panel data in Chapter 3 (9.0%) can be explained by considering the differing response variables of the two questions. The ECHP question contained a binary-response (Yes/No) question regarding the ability to adequately heat the home. It is believed that a large portion of the 'occasionally' fuel-poor (highlighted in the national household survey) are not declaring problems of fuel poverty in the ECHP data from the 1990s which appears to capture *persistent*, as opposed to *intermittent*, fuel poverty and, as such, a lower incidence is reported using the binary-response (ECHP) variable. Fuel-poverty estimates of 8.0%, 5.9%, 6.5% and 5.1% were produced for the years 1994-97 inclusive using the ECHP. If the continuing downward trend persisted over the years 1998-2000, then it is suggested that the chronic estimate of fuel poverty of 4.7% reported in the national household survey corresponds well with the results of the ECHP in Chapter 3.

It is suggested that much of this decline in chronic levels of fuel poverty is related with the rising living standards witnessed in Ireland over this ('Celtic Tiger') boom period of 1994-2001 (Figure 11.3). Increasing PPP-adjusted GDP per capita for the years 1994-2001 is almost perfectly correlated with the reduced levels of fuel poverty over those years (R=-0.97, P<0.001). The slowdown of the Irish economy during 2002 implies that further reductions in the level of fuel poverty are unlikely to be realised unless targeted State subvention aimed at improving the thermal standards of the housing stock occurs over the coming years.

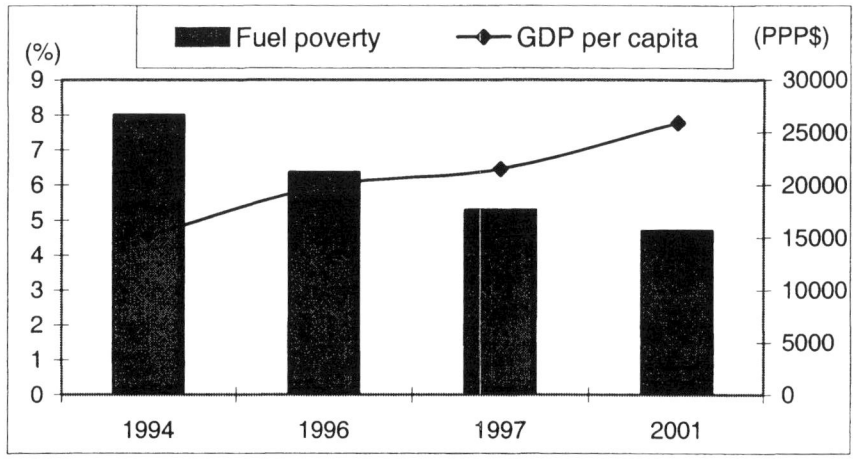

Figure 11.3 Chronic fuel poverty in Ireland, 1994-2001 (% of households)

Results from the national household survey show that the highest incidence of fuel poverty is found among the long-term ill and disabled, where 44.8% of such households (11,000 in absolute numbers) are demonstrating fuel poverty. However, this result is based on a relatively small sample of households and should be treated with care. Lone-parent households are identifiable as the second-highest group of fuel-poverty sufferers in Ireland, with over two-fifths (40.2%) declaring some inability to heat the home adequately (29,000 households). Income is closely associated with fuel poverty, and the results of this study bear out this tenet; 35.6% of households under €12,700 (IR£10,000) are suffering (representing some 121,000 households). A similarly strong result is found with social class, where 34.6% of those in the lowest group (E) are reporting fuel poverty (34,000 households). Other groups with high incidences include: local-authority tenants (33.8% or 57,000 households), the unemployed (30.5% or 16,000 households), one-person occupied households (28.4% or 51,000 households), those separated, divorced or widowed (28.2% or 37,000 households), lone female pensioners (28.1% or 11,000 households) and those who completed their education at primary level (25.6% or 47,000 households). Chronic fuel poverty is proportionately highest among households with four or more dependent children, where 55.5% of fuel poverty appears to be persistent, followed by 44.1% of fuel-poor private tenants and 42% of students living in fuel poverty.

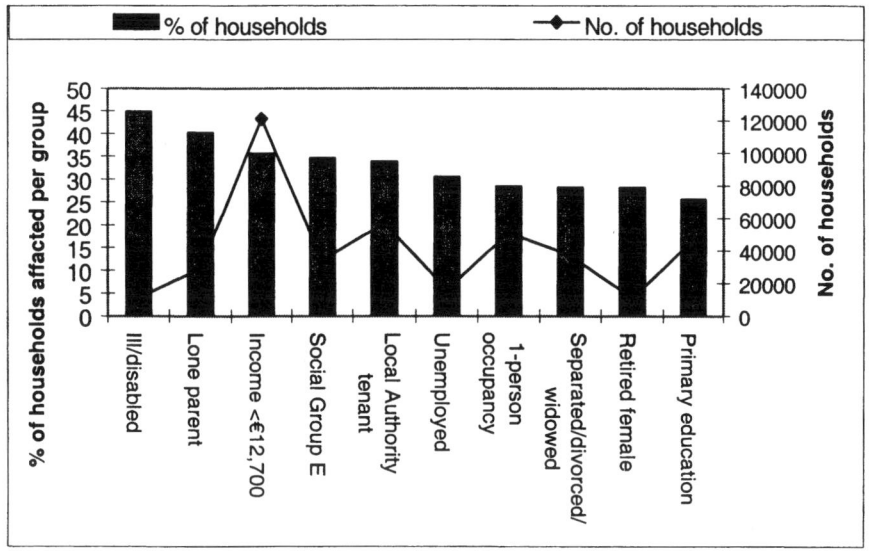

Figure 11.4 Fuel poverty in Ireland: 10 risk groups

The results of the national survey demonstrated a strong association between fuel poverty and household condensation and an even stronger association with household damp, where the incidence of fuel poverty is almost four times that found among households without damp spores. The findings of the survey also indicated that the State fuel allowance is a necessary but insufficient measure in tackling fuel poverty. Households claiming the fuel subsidy report an incidence of fuel poverty some three times higher than non-claiming households. However, the fuel allowance appears to impact positively on the *severity of experience*, significantly reducing the proportion of chronic fuel-poor households.

The chapter also presented the results of the survey regarding the key reasons why energy-saving measures are not adopted by households. The findings indicate a large 'information gap' in the market of domestic energy-efficiency measures, with over a half of households either unaware of the existence or unaware of the benefits of energy-saving measures in the home. In addition, over a third of households identified financial constraints to retrofitting, with only a very small proportion blaming transactions' costs.

Housing Deprivation and Health

Chapter 5 presented an analysis of the association between health and a variety of social indicators of housing deprivation in Europe using the ECHP database on social indicators. This quantitative analysis showed that there are a number of apparent serious causes for concern. The first major conclusion pertains to the relationship found between housing conditions and health. A number of social indicators relating to housing deprivation were employed to assess the health effects of lacking various socially perceived necessities. The results show that, with the exception of overcrowding (for which no relationship is found), all other (16) indicators are found to be significantly associated with increased incidence of poor health status in EU-14. The (in)ability to adequately heat the home – a key socially perceived necessity and indicator of fuel poverty – is found to result in higher incidences of poor health status in the Member States analysed, with over twice the level of poor health reported among fuel-poor households. However, results are far from conclusive with regard to the precise effect on health of fuel poverty *alone* . It was also found that the least energy efficient housing stocks tend to suffer from the highest levels of poor health. Generally, the highest levels of aggregate poor health were found in southern Europe, where housing conditions (regarding both thermal efficiency and general housing deprivation) appear to be the poorest. In northern Europe, Austria, France, Luxembourg and the UK demonstrate high incidences of poor health among deprived households and those with below-par housing conditions.

Housing affordability is also found to be related to health status in EU-14, with significantly increased levels of poor health reported among those with financially burdensome housing costs and those unable to pay utility bills on time. In addition, poor health is found to be highly associated with housing dissatisfaction, with 30.5% of dissatisfied households in EU-14 declaring poor or very poor health; the incidence of poor health of 69.7% of dissatisfied households in Portugal is especially worrying.

It is important to bear in mind that, while the incidence of poor health is important across EU-14 in identifying aggregate health outcomes, it is perhaps more useful to compare the variations in the incidence of poor health associated with various social indicators, as this indicates more precisely the health effects of a given deprivation indicator. This is because poorer countries (like Greece, Spain and Portugal) are likely to report higher incidences of poor health regardless of whether or not they declare deprivation indicators. Some distinct patterns emerge when proportionate variations in health are examined across EU-14. Ireland appears to suffer from the largest proportionate increases in poor health among households demonstrating an indicator of housing deprivation; Greece and Italy also suffer from large proportionate increases in poor health with regard to housing deprivation and fuel poverty. The Netherlands suffers persistently from the largest proportionate increases in poor health with regard to housing affordability and satisfaction with housing, while Luxembourg also performs poorly with regard to the former.

Although many results appear to corroborate the majority of research regarding the relationship between housing deprivation and health, there are some apparent anomalies. Despite having among the least energy efficient housing stocks in northern Europe and among the highest levels of income poverty and inequality in Europe, Irish households persistently report among the lowest absolute levels of poor health in EU-14. A number of interpretations are possible.

However, one plausible explanation is that the Irish, either intentionally or otherwise, 'under-declare' their self-perceived levels of poor health. This may be due to cultural nuances. Demography may also play a part; countries with relatively young populations are more likely to report good health status than ageing populations. Cultural idiosyncrasies always play a part in self-reported results, and it is possible that they affect the (relatively low) reported magnitude of absolute levels of poor health in Ireland. The findings regarding the lack of a relationship between health and overcrowding also require some additional comments. There is a considerable medical and social-policy literature highlighting the effects on health of crowding in the home. However, much of this research uses objective measures of overcrowding (as opposed to subjective, self-reporting measures). It is possible that the results are showing some interplay across EU-14 owing to cultural factors. Many households may feel embarrassed to declare shortage of space, while others may not regard their housing conditions as crowded when, using an objective measurement, such households may be defined as objectively overcrowded.[4]

Fuel Poverty and Health in Ireland

The results from the national household survey regarding the health impacts of fuel poverty in Ireland, presented in Chapter 6, provide inconclusive evidence and do not support the relatively small body of literature demonstrating a link between poor housing and adverse health impacts. Any link with adverse health outcomes demonstrated in this chapter is far from conclusive and no causal relationships can be identified. As with all self-reported data, it is important to stress that cultural idiosyncrasies often play a part in survey results. It was shown in Chapter 2, for instance, that Ireland and Italy continuously under-declared their perceived level of poor health, even when it was shown that generalised health outcomes would have been poorer using objective (non-self-reported) measures. It is likely in this study that much of the results are actually conservative, i.e. some households, for a variety of reasons, may not wish to state, for example, that they are suffering from poor emotional or mental health. It is important to take this on board when analysing the data. In this regard, some apparently moderate findings may take on more weighty significance.

Nonetheless, some key conclusions can be made. Fuel-poor households report lower levels of health status (and higher levels of poor or impaired health). There also appears to be something of a dynamic aspect to the results, for it was demonstrated that the proportion of fuel-poor households believing their health status is impaired now compared to a year ago is over twice the incidence reported amongst all other households. However, (once again) these results carry caveats, as discussed in the chapter.

---

[4] In the cross-country analysis of housing conditions in Europe (Chapter 4), it was reported that Irish and Italian households continuously under-declare their perceived levels of household crowding, despite objective measures of overcrowding reporting otherwise.

While the (older) age profile of the fuel-poor is likely to play a part in some results pertaining to health impacts, the data demonstrate increased levels of hardship and adverse health outcomes among the fuel-poor, though some activities were found to be far from significant. All four manifestations of poor health were found to be associated with the fuel-poverty status of the household, and such manifestations were found to result in similar results for both physical and mental/emotional health. With regard to personal feelings, the ability to feel full of life, calm and peaceful, energetic and not lethargic were all found to be negatively associated with the incidence of fuel poverty.

It was interesting to note that many fuel-poor households are self-aware of their health risks and reduced health status. More households living in fuel poverty believe they become ill more easily than on average, they expect their health to disimprove over time, they do not believe they are generally as healthy as the typical person, and they are far less likely to state that their health is excellent. Using objective measures of health status, it was shown that the fuel-poor are more likely to visit their GP regularly. They also demonstrated relationships between other objective health outcomes (e.g. they are more likely to be admitted to hospital as a day-case or to A&E or as an outpatient). Interestingly, fuel-poor households are far more likely to seek the advice of a pharmacist than other households. The key adverse outcome found here relates to the result that fuel-poor households are almost twice as likely to suffer from a chronic illness or disease than other households, with one-in-five affected. The variation is most acute for chronic diseases of the respiratory system and long-term depression.

An interesting finding of this part of the study relates to housing as a self-perceived causal factor in the levels of poor health status, or more particularly in the levels of chronic diseases. Fuel-poor households are relatively far more likely (three times, in fact) to blame housing conditions as a key cause for their illness than other households. It could be argued that this form of suggestive questioning, based on the UK SF36, may elicit somewhat biased results. However, as many people are unaware of the link between housing and ill health, so the result is likely to be erring on the conservative side.

Fuel-poor households worry more than other households. In particular, they worry more about their finances and job security, their children's health, and their own physical health. They do *not*, however, worry to a significantly higher degree about their housing situation. As might be expected, quality of life and general life satisfaction is found to be negatively affected by the presence of fuel poverty. Fuel-poor households are more likely to report both poor quality of life and low levels of overall life satisfaction, with proportionate variations of up to 200% between households suffering from fuel poverty and those not enduring such problems.

Risk Factors and Fuel-Poor Households

Data from the national household survey were employed to conduct a risk-factor analysis of fuel-poor households. It appears from the analysis presented in Chapter 7 that fuel-poor households not only exhibit high levels of endogenous risk factors pertaining to indoor cold stress, but they also exhibit high levels of a range of other, exogenous, generalised risk factors associated with impaired health status. In this regard, policy-makers are faced with a multi-faceted job in mitigating fuel poverty-associated ill health.

It was demonstrated that, while the duration of time spent outdoors during winter did not appear to exhibit a strong relationship with the fuel-poverty status of a household, the presence and duration of shivering outdoors was a more telling variable. Such a physiological outcome measure indicates inadequate thermal protection from the cold, and fuel-poor households were shown to suffer more, and for longer periods of time, from spells of cold-induced shivering. The results were dramatic when the question was replicated to elicit information regarding households' indoor environment. Some 56.5% of fuel-poor households suffer from episodes of shivering indoors on a typical, cold winter night, compared with just 15.8% of those unafflicted by fuel poverty. Such findings are very similar to those found in previous epidemiological studies on the adverse health effects of cold stress.

However, when the risk-factor analysis examined other, exogenous (non-cold) hazards associated with impaired health, fuel-poor households were found to fare worse than other households. Specifically, fuel-poor households smoke more, exercise less and eat less healthily than other households. These findings suggest that socio-economics also play a large part in health inequalities associated with the presence of fuel poverty. This is not a particularly radical result; Kunst *et al.* (1991) showed that the reduction in excess winter deaths in the Netherlands was closely related to factors associated with socio-economic progress, and Townsend has always stressed the importance of the class structure in his materialist explanation of health divides. A logical conclusion from this research is that typical State-backed energy-efficiency retrofit programmes aimed at low-income households are unlikely to result in the alleviation of all ill health related to fuel poverty, as it appears from this study that fuel-poor households, like those living in generalised income poverty and deprivation, suffer from a plateau of other health hazards associated with being less well-off.

Fuel Poverty, Thermal Comfort and Occupancy

The results of the national household survey indicate that there are large numbers of fuel-poor households living in thermal discomfort. Self-reported measures of thermal comfort demonstrate that about one-in-ten households (which represents 130,000 homes) nationwide report thermal discomfort, with higher levels in kitchens and bedrooms, while fuel-poor households are up to five times more likely to report thermal discomfort in certain rooms in their homes, such as the master bedroom and the bathroom. Furthermore, over half of fuel-poor households endure shivering episodes on typical cold winter evenings; such physiological responses to cold temperatures indicate cold strain, reduced resistance to respiratory infections and potential cardiovascular stress. When more objective output measures are employed, a similar pattern of thermal discomfort is demonstrated, with those enduring fuel poverty far more likely to be inhabiting homes with living-room temperatures below those set by the World Health Organisation as minimum levels of warmth to sustain resistance to impaired health status. Overall, three-in-ten fuel-poor households live in dwellings where the living room is heated to levels which can result in adverse health impacts, even on the young and healthy. Not all fuel-poor households are reporting (self-perceived) thermal discomfort. It appears from the survey that many fuel-poverty sufferers in Ireland demonstrate occasional difficulties in achieving affordable home heating, while chronic sufferers appear to endure thermal discomfort and the coldest housing.

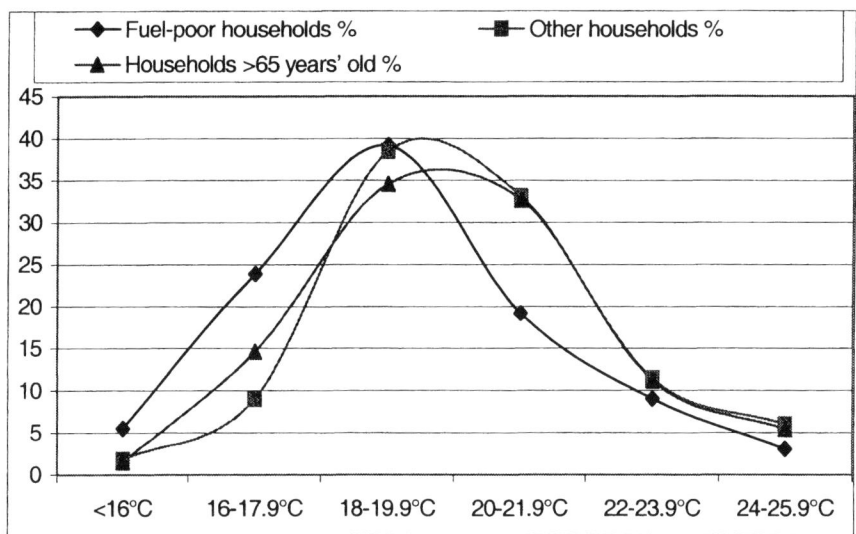

Figure 11.5 **Living-room temperatures of fuel-poor, elderly and all other households (% households, Ireland, 2001)**

The age profile of those living in thermally uncomfortable housing is notable. Substantial portions of the over-65s are enduring cold housing. Using self-reported data, over a quarter of those over 65 are declaring thermal discomfort in some rooms in their home. Furthermore, about 30% of the elderly report episodic shivering which may result in cardiovascular strain in older people. However, the results based on the objective data on ambient living-room temperature make for more alarming reading, with over a half of old-age pensioners living in housing where living-room temperatures are below those set by the WHO as minimum satisfactory levels of warmth for the vulnerable. Socio-economic analysis demonstrates that those in poorer social groups are far more likely to live in cold housing, as are those with large numbers of dependent children. Such a finding is worrying when it is considered that young children generally require warmer ambient conditions than healthy adults. Thus, many young children may be living in under-heated homes, exposing themselves at a young age to health risks which may reduce resistance to infection.

The existence of fuel poverty also plays a part on household occupancy. One-third of fuel-poor households inhabit cold, unheated rooms during winter. In addition, a small proportion of households nationally do not use rooms because they are too cold to inhabit during the winter. Approximately one-fifth of households occupied by older people follow these occupancy patterns which may be considered manifestations of fuel poverty.

Fuel Poverty and Excess Winter Mortality

Using data from 1988-97, relative excess winter mortality is found to be highest in southern Europe, Ireland and the UK, where seasonality in mortality of 18-28% is calculated. The results of this analysis, presented in Chapter 9, also show how Scandinavian and other northern-European countries are relatively unaffected by the problem. Such results are startling, especially for southern Europe where virtually no published work exists regarding seasonal variations in mortality. Paradoxically, countries with the mildest winter climates, where mean environmental temperatures remain above 5°C, exhibit the highest variations in seasonal mortality. Other climatic variables, such as mean winter precipitation levels and relative humidity, are not significantly related to cross-country variations in excess winter deaths. However, the strong relationship demonstrated between mean cross-country winter environmental temperatures and levels of relative excess winter mortality indicates that certain populations are far more vulnerable to cold exposure than others.

Moreover, available data on cross-country thermal-efficiency standards in housing indicate that those countries with the poorest housing (Portugal, Greece, Ireland, the UK) demonstrate the highest excess winter mortality. If the ability of a population to protect themselves from cold spells is a key factor in such pronounced seasonality in southern and western Europe, as has been mentioned in previous literature, then it would appear that improving the thermal standards of housing could be an effective preventative intervention in curbing excess deaths. Such a health strategy would also assist in the alleviation of fuel poverty which, this study shows, is also highest in those countries in southern and western Europe with the poorest energy efficiency.

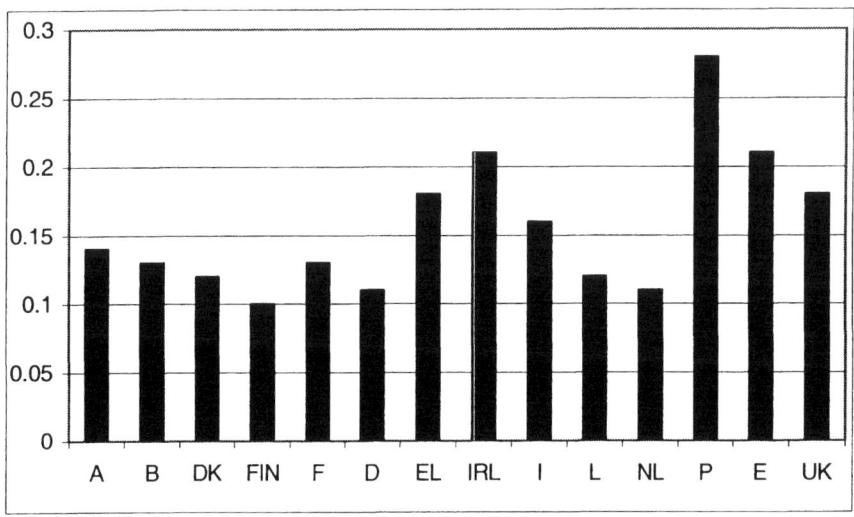

Figure 11.6 Relative excess winter mortality in EU-14 (mean, 1988-97)

Socio-economic indicators of well-being (poverty, income inequality, deprivation and fuel poverty) are also associated with cross-country levels of excess winter mortality. This suggests that levels of excess winter mortality could be reduced through socio-economic progress, as was found in a longitudinal analysis of the Netherlands, especially in countries with more unequal income distribution. Macroeconomic data indicate that levels of relative excess winter mortality are associated with per-capita GNP, but not State expenditure on education. Lifestyle risk factors are also not associated with seasonal variations in mortality, unlike all-year mortality rates. However, a number of indicators of healthcare provision are significantly associated with cross-country variations in seasonal mortality. Strong associations are reported for public health expenditure as a proportion of per-capita GNP, purchasing-power adjusted health expenditure per capita and hospital-bed ratios. The latter indicator is especially worrying, as it indicates that potential bed shortages could be resulting in increased seasonal mortality in winter when resources are at their most stretched. Such findings are cause for concern for those countries with relatively low hospital bed ratios and short average durations of stay, most notably Ireland and Spain.

Although this research has not proven causality conclusively, the strong, positive relationship with environmental temperature and the equally strong associations with thermal standards in housing indicate that improving the thermal efficiency of housing in southern and western Europe could play a strong role in reducing the large seasonal variations in mortality found in these countries.

Summary and Conclusions

A number of hypotheses were outlined in Chapter 1. The results of the study are now summarised under these headings.

1. *Fuel poverty and poor domestic energy inefficiency is strongly associated with impaired health status.*

    Across Europe, higher levels of poor health are reported among fuel-poor households. However, close examination of Irish survey data reveal inconclusive evidence of adverse health outcomes. Although fuel-poor households are more likely to suffer ill health, it cannot be shown that it is fuel poverty, *per se*, that is the causal factor in such ill health.

2. *Fuel poverty and poor domestic energy inefficiency is strongly associated with high levels of excess winter mortality.*

    This statement appears to be correct. A strong, significant correlation is found between levels of excess winter deaths and levels of fuel poverty in Europe. Countries with high fuel poverty invariably have high levels of excess winter mortality and *vice versa*.

3. *A Consensual approach to calculating fuel poverty results in more conservative, but more complete, estimates than those resulting from the standard definition.*

    Using a standard definition of fuel poverty (households that spend more than 10% of income on home-heating costs), a rate of 20.7% is found using 1995 data, while a rate of 9.0%-12.4% is found using the Consensual indicators. Similar variations are found for the UK (16.4% using a standard definition and 9.3%-12.5%). The standard definition is wholly arbitrary in design and it is argued in this study that the composite measure of fuel poverty is a more complete, albeit conservative and imperfect, estimate. This is because it is based on experiential indicators and socially perceived necessities, as opposed to more arbitrary objective measurements based on what 'experts' deem to be most appropriate.

4. *The fuel-poor exhibit higher risk factors associated with poor health.*

    This postulation is found to be very much palpable using the data reported in this study. The fuel-poor are less likely to protect themselves from the cold indoors *and* outdoors. They suffer from poorer diets, take part less in physical exercise and smoke more than other (non-fuel-poor) households.

5. *The fuel-poor endure reduced thermal comfort in the home and lower ambient household temperatures.*

   The survey data indicate that this is very true. Almost a third of the fuel-poor live in cold housing with a living-room temperature below that recommended by the World Health Organisation; just one-in-ten other households endure such temperatures. The fuel-poor also endure higher levels of self-reported thermal discomfort than the fuel-rich. Longer incidences of cold strain are also found among the fuel-poor. The national household survey also indicated that over half the elderly endure living-room temperatures below those recommended by the WHO as minimum requirements for remaining healthy. These results are particularly alarming in terms of health implications.

6. *Seasonal variations in mortality are a product of a wide range of social and economic factors, not just fuel poverty and inadequately insulated homes.*

   This appears to be a legitimate supposition. The strongest cross-country relationships with seasonal mortality are found with environmental temperature, per-capita GDP, healthcare provision, levels of income poverty, inequality, deprivation and fuel poverty, and domestic thermal efficiency. It appears from the data that excess winter mortality could be reduced largely through socio-economic progress resulting in reduced poverty and income inequality (especially in countries with more unequal income distribution) and improvements in domestic energy efficiency resulting in lower fuel poverty. Lifestyle risk factors are not associated with seasonal variations in mortality, unlike all-year mortality rates.

7. *Domestic energy-efficiency standards in Ireland and the UK are among the poorest in Europe.*

   Irish and British energy-efficiency standards are below-par relative to their *northern-European* counterparts, although available data suggest that those in southern Europe (Portugal, Greece, Spain and Italy) are considerably worse. Irish housing standards are certainly among the worst in northern Europe, with the UK, Austria and Belgium similarly under-protected from the external environment.

8. *Fuel poverty in Ireland and the UK is among the highest in Europe (1994-97).*

   This statement is found to be false. One of the most unexpected results of the research was the fuel-poverty findings in Mediterranean nations. Southern-European nations such as Portugal, Spain, Greece and Italy all exhibit the highest levels of national fuel poverty. However, Belgium, France, the UK and Ireland all demonstrate relatively high levels of fuel poverty in northern Europe. Irish and British fuel poverty is particularly concentrated in poorer social groups such as lone parents, lone pensioners and local-authority tenants.

9. *Health impairment associated with fuel poverty is among the highest in Ireland.*

   This proposition is accepted. Although ECHP data suggest that Irish households under-declare their levels of ill-health and over-estimate their health status in general, variations in impaired health associated with being fuel-poor are among the highest in Ireland for all indicators of fuel poverty analysed in the study. The UK also fares poorly in this respect.

10. *Excess winter mortality in Ireland and the UK is among the highest in Europe.*

    Portugal is found to have the highest level of excess winter mortality. Ireland and Spain follow with a rate of 21% which is over twice that found in more northerly, colder climes. The UK's rate of 18% follows closely. As such, the hypothesis that Ireland and the UK suffer from among the highest levels of seasonal mortality is found to be true.

11. *The information gap is the main reason for market failure in domestic energy efficiency.*

    This is true. Over 50% of households lacking various energy-saving measures are either unaware of the benefits of installing such energy-saving technology or do not know of the existence of these measures. The other main blockage in the market for domestic energy efficiency relates to lack of income and more pressing priorities for funds.

Final Word

Using cross-country panel data, this research has demonstrated that there appears to be a strong relationship between domestic energy inefficiency, fuel poverty and adverse human health outcomes, most acutely the level of excess winter mortality. However, national incidences of excess winter deaths are found to be strongly related to a range of variables, not just housing conditions and fuel poverty. While rising incomes appear to have impacted positively on the level of fuel poverty in Ireland and the UK over the past decade, it is highly likely that improved State subvention will be required in order to undertake remedial work on the capital stock in an effort to reduce fuel poverty most efficiently. Current (as opposed to capital) expenditure on fuel poverty (e.g. a fuel allowance) reduces the severity of experience of fuel poverty, but appears to do little to reduce the overall incidence. In the absence of capital investment, the national level of fuel poverty is likely to fluctuate broadly in line with the macroeconomic environment.

APPENDIX I

Table 1 Ability to heat home adequately: results of a Probit model for 1994 and 1995

Dependent Variable: HEAT HOME			1994				1995		
Explanatory Variables		Coeff	t Stat	P-Value	Marginal Effects	Coeff	t Stat	P-Value	Marginal Effects
Constant		2.512	80.83	-	-	2.261	72.71	-	-
Age Group	16-34	-.328	-17.43	[.000]***	-9.1%	-.359	-19.21	[.000]***	-9.7%
	35-44	-.184	-9.94	[.000]***	-5.1%	-.181	-9.70	[.000]***	-4.9%
	45-54	-.026	-1.53	[.124]	-0.7%	-.018	-1.04	[.296]	-0.5%
	55-64	-.078	-5.01	[.000]***	-2.1%	-.044	-2.81	[.005]**	-1.2%
Household Composition	Number of Adults	-.145	-40.93	[.000]***	-4%	-.115	-33.21	[.000]***	-3.1%
	Number of Children	-.054	-10.78	[.000]***	-1.5%	-.031	-6.17	[.000]***	-0.8%
Marital Status	Separated	-.298	-8.32	[.000]***	-8.3%	-.306	-8.37	[.000]***	-8.3%
	Divorced	-.025	-.96	[.333]	-0.7%	.068	2.57	[.010]**	1.8%
	Widowed	-.149	-9.07	[.000]***	-4.1%	-.128	-7.77	[.000]***	-3.4%
	Single	-.106	-8.34	[.000]***	-2.9%	-.040	-3.14	[.002]**	-1%
Income	Self employed	-.094	-7.23	[.000]***	-2.6%	-.167	-12.94	[.000]***	-4.5%
	Pension	-.183	-13.62	[.000]***	-5.1%	-.178	-13.24	[.000]***	-4.8%
	Unemployed	-.267	-10.66	[.000]***	-7.5%	-.231	-8.29	[.000]***	-6.2%
	Other benefit	-.249	-13.47	[.000]***	-6.9%	-.188	-9.81	[.000]***	-5.1%
	Private Income	-.147	-4.99	[.000]***	-4.1%	-.291	-9.47	[.000]***	-7.9%

Tenure	Tenant	.024	2.34	[.019]**	0.6%	-.040	-3.90	[.000]***	-1.1%
	Rent free	-.499	-26.45	[.000]***	-13.9%	-.099	-4.76	[.000]***	-2.7%
Accommodation type	Semi-detached	.074	6.98	[.000]***	2%	.128	11.53	[.000]***	3.4%
	Terraced								
	Apartment Small	-.150	-12.58	[.000]***	-4.2%	.278	21.32	[.000]***	7.5%
	Apartment Large	-.091	-7.50	[.000]***	-2.5%	.023	2.02	[.043]**	0.6%
	Other Accommodation	.066	2.13	[.033]**	1.8%	.118	5.26	[.000]***	3.2%
Health Status	Good	-.124	-11.54	[.000]***	-3.4%	-.076	-7.02	[.000]***	-2%
	Fair	-.281	-22.08	[.000]***	-7.8%	-.264	-20.57	[.000]***	-7.1%
	Bad	-.793	-46.44	[.000]***	-22.2%	-.738	-42.60	[.000]***	-20.1%
	Very Bad	-.740	-29.12	[.000]***	-20.7%	-.683	-25.13	[.000]***	-18.6%
Education	Secondary finished	-.077	-5.59	[.000]***	-2.1%	.016	1.16	[.242]	0.4%
	Secondary not finished	-.850	-68.20	[.000]***	-23.8%	-.768	-60.63	[.000]***	-20.9%
Housing Allowance		-.399	-19.93	[.000]***	-11.1%	-.396	-19.05	[.000]***	-10.7%
Number of Observations		128045				127912			
Log Likelihood		-63773.5				-61995.9			
Mean of Dependent Variable		.740708				.761336			
Percentage Correct Prediction		75.7%				77%			
Pseudo R^2		.147070				.128931			

***significant at 1% level; **significant at 5% level.

Table 2 Ability to heat home adequately: results of a Probit model for 1996 and 1997

Dependent Variable: HEAT HOME			1996				1997		
Explanatory Variables		Coeff	t Stat	P-Value	Marginal Effects	Coeff	t Stat	P-Value	Marginal Effects
Constant		2.151	68.04	-	-	.844	33.60	-	-
Age Group	16-34	-.348	-18.22	[.000]***	-8.9%	-.197	-11.60	[.000]***	-6.8%
	35-44	-.180	-9.49	[.000]***	-4.6%	-.060	-3.57	[.000]***	-2.1%
	45-54	-.889	-.49	[.618]	-0.2%	-.035	-2.25	[.024]**	-1.2%
	55-64	-.010	-.62	[.532]	-0.2%	-.077	-5.38	[.000]***	-2.7%
Household Composition	Number of Adults	-.121	-34.08	[.000]***	-3.1%	-.012	-3.86	[.000]***	-0.4%
	Number of Children	-.027	-5.21	[.000]***	-0.7%	-.017	-3.83	[.000]***	-0.5%
Marital Status	Separated	-.246	-6.30	[.000]***	-6.3%	-.068	-2.07	[.038]**	-2.3%
	Divorced	.048	1.86	[.062]**	1.2%	.088	4.57	[.000]***	3%
	Widowed	-.149	-8.93	[.000]***	-3.8%	.050	3.29	[.001]***	1.7%
	Single	.729	.56	[.573]	0.1%	.125	11.56	[.000]***	4.3%
Income	Self employed	-.128	-9.71	[.000]***	-3.3%	.189	15.12	[.000]***	6.5%
	Pension	-.191	-14.09	[.000]***	-4.9%	-.059	-4.79	[.000]***	-2%
	Unemployed	-.239	-8.19	[.000]***	-6.1%	-.169	-6.83	[.000]***	-5.9%
	Other benefit	-.206	-10.66	[.000]***	-5.3%	.111	6.53	[.000]***	3.8%
	Private Income	-.084	-2.49	[.013]**	-2.1%	.195	7.03	[.000]***	6.8%
Tenure	Tenant	.106	9.88	[.000]***	2.7%	-.336	-36.17	[.000]***	-11.6%
	Rent free	-.094	-4.46	[.000]***	-2.4%	-.176	-9.69	[.000]***	-6.1%

Accommodation type	Rent free	-.094	-4.46	[.000]***	-2.4%	-.176	-9.69	[.000]***	-6.1%
	Semi-detached	.359	31.50	[.000]***	9.2%	.273	28.72	[.000]***	9.5%
	Terraced								
	Apartment Small	.422	31.79	[.000]***	10.8%	-.039	-3.59	[.000]***	-1.3%
	Apartment Large	.045	3.93	[.000]***	1.1%	.149	13.90	[.000]***	5.1%
	Other Accommodation	.613	23.43	[.000]***	15.8%	-.661	-34.72	[.000]***	-23%
Health Status	Good	-.034	-3.13	[.002]**	-0.8%	.132	14.71	[.000]***	4.6%
	Fair	-.181	-13.87	[.000]***	-4.6%	-.058	-5.49	[.000]***	-2%
	Bad	-.658	-37.36	[.000]***	-16.9%	-.456	-29.65	[.000]***	-15.8%
	Very Bad	-.653	-23.38	[.000]***	-16.8%	-.452	-17.84	[.000]***	-15.7%
Education	Secondary finished	-.028	-1.99	[.046]**	-0.7%	-.065	-6.48	[.000]***	-2.2%
	Secondary not finished	-.786	-60.65	[.000]***	-20.2%	-.298	-30.89	[.000]***	-10.3%
Housing allowance		-.385	-18.35	[.000]***	-9.9%	-.159	-9.81	[.000]***	-5.5%
Number of Observations		130639				138130			
Log Likelihood		-60067.0				-84430.0			
Mean of Dependent Variable		.778542				.655578			
Percentage Correct Prediction		78.8%				67.4%			
Pseudo R^2		.137743				.064895			

***significant at 1% level; **significant at 5% level.

Table 3 Ability to pay utility bills in past year: results of a Probit model for 1994 and 1995

Dependent Variable: BILLS			1994				1995		
Explanatory Variables		Coeff	t Stat	P-Value	Marginal Effects	Coeff	t Stat	P-Value	Marginal Effects
Constant		-2.09	-57.48	-	-	-1.761	-45.79	-	-
Age Group	16-34	.333	13.33	[.000]***	4.8%	.299	11.20	[.000]***	3.5%
	35-44	.224	9.06	[.000]***	3.2%	.209	7.94	[.000]***	2.5%
	45-54	.228	9.70	[.000]***	3.3%	.186	7.32	[.000]***	2.2%
	55-64	.172	8.04	[.000]***	2.5%	.060	2.57	[.010]**	0.7%2
Household Composition	Number of Adults	.080	18.06	[.000]***	1.1%	.047	9.91	[.000]***	0.5%
	Number of Children	.121	20.78	[.000]***	1.7%	.104	16.87	[.000]***	1.2%
Marital Status	Separated	.345	8.54	[.000]***	5%	.392	9.073	[.000]***	4.6%
	Divorced	.234	8.23	[.000]***	3.4%	.220	7.23	[.000]***	2.6%
	Widowed	.142	6.38	[.000]***	2%	.101	4.17	[.000]***	1.2%
	Single	.048	3.01	[.003]**	0.7%	.038	2.21	[.027]**	0.4%
Income	Self employed	.256	15.79	[.000]***	3.7%	.248	14.56	[.000]***	2.9%
	Pension	.197	10.91	[.000]***	2.8%	.133	6.78	[.000]***	1.5%
	Unemployed	.381	13.31	[.000]***	5.5%	.529	16.69	[.000]***	6.3%
	Other benefit	.296	13.80	[.000]***	4.3%	.308	13.10	[.000]***	36%
	Private Income	.357	10.24	[.000]***	5.2%	.452	12.11	[.000]***	5.3%
Tenure	Tenant	.186	14.28	[.000]***	2.7%	-.107	-7.14	[.000]***	-1.2%
	Rent free	.115	4.53	[.000]***	1.6%	.376	15.37	[.000]***	4.4%

Category	Variable	Coef	t	p-value	%	Coef	t	p-value	%
Accommodation type	Semi-detached	.096	6.89	[.000]***	1.4%	.034	2.33	[.020]**	0.4%
	Terraced								
	Apartment Small	.208	13.40	[.000]***	3%	-.029	-1.7	[.081]*	-0.3%
	Apartment Large	.144	9.07	[.000]***	2.1%	-.167	-9.70	[.000]***	-1.9%
	Other Accommodation	-.137	-3.07	[.002]**	-2%	-.425	-10.89	[.000]***	-5%
Health Status	Good	-.170	-12.73	[.000]***	-2.4%	-.198	-13.80	[.000]***	-2.3%
	Fair	-.021	-1.34	[.179]	-0.3%	-.045	-2.67	[.008]**	-0.5%
	Bad	.140	6.42	[.000]***	2%	.109	4.55607	[.000]***	1.3%
	Very Bad	.298	9.65	[.000]***	4.3%	.293	8.40	[.000]***	3.5%
Education	Secondary finished	.090	5.35	[.000]***	1.3%	.037	2.00	[.045]**	0.4%
	Secondary not finished	.327	20.65	[.000]***	4.7%	.284	16.57	[.000]***	3.3%
Housing allowance		-.289	-14.21	[.000]***	-4.2%	-.365	-16.87	[.000]***	-4.3%
Number of Observations		128045				127912			
Log Likelihood		-34807.6				-28796.8			
Mean of Dependent Variable		.084267				.064685			
Percentage Correct Prediction		91.5%				93.5%			
Pseudo R^2		.034908				.029552			

***significant at 1% level; **significant at 5% level.

Table 4 Ability to pay utility bills in the past year: results of a Probit model for 1996 and 1997

Dependent Variable: BILLS		1996				1997			
Explanatory Variables		Coeff	t Stat	P-Value	Marginal Effects	Coeff	t Stat	P-Value	Marginal Effects
Constant		-2.219	-48.89	-	-	-2.189	-55.85	-	-
Age Group	16-34	.577	16.85	[.000]***	4.7%	.236	8.42	[.000]***	2.4%
	35-44	.478	14.20	[.000]***	3.9%	.207	7.53	[.000]***	2.1%
	45-54	.4450	13.61	[.000]***	3.6%	.167	6.38	[.000]***	1.7%
	55-64	.216	6.82	[.000]***	1.7%	.091	3.72	[.010]**	0.9%
Household Composition	Number of Adults	.042	7.89	[.000]***	0.3%	.084	16.92	[.000]***	0.8%
	Number of Children	.1341	20.17	[.000]***	1.1%	.118	18.78	[.000]***	1.2%
Marital Status	Separated	.411	8.26	[.000]***	3.3%	.208	4.35	[.000]***	2.1%
	Divorced	.336	10.82	[.000]***	2.7%	.156	5.18	[.000]***	1.6%
	Widowed	.127	3.93	[.000]***	1%	.162	6.48	[.000]***	1.6%
	Single	.119	6.21	[.000]***	0.9%	.047	2.71	[.027]**	0.4%
Income	Self employed	-.020	-.91	[.360]	-0.1%	.277	15.63	[.000]***	2.8%
	Pension	.039	1.62	[.104]	0.3%	.146	7.23	[.000]***	1.5%
	Unemployed	.638	19.32	[.000]***	5.2%	.445	13.54	[.000]***	4.6%
	Other benefit	.365	15.15	[.000]***	3%	.221	9.18	[.000]***	2.3%
	Private Income	-.056	-.94	[.343]	-0.4%	.286	7.10	[.000]***	2.9%
Tenure	Tenant	-.053	-3.25	[.001]**	-0.4%	.075	5.08	[.000]***	0.7%
	Rent free	.023	.68	[.495]	0.1%	-.093	-2.95	[.000]***	-0.9%
Accommodation type	Semi-detached	-.016	-.89	[.373]	-0.1%	.811	.52	[.020]**	0.0%
	Terraced								

		Coef.	t	p-value	Marg.	Coef.	t	p-value	Marg.
	Apartment Small	.126	6.44	[.000]***	1%	.214	12.68	[.081]*	2.2%
	Apartment Large	.071	3.77	[.000]***	0.5%	.121	7.06	[.000]***	1.2%
	Other Accommodation	.170	5.20	[.000]***	1.4%	-.255	-6.42	[.000]***	-2.6%
Health Status	Good	.042	2.48	[.013]**	0.3%	-.174	-12.05	[.000]***	-1.8%
	Fair	.277	13.80	[.000]***	2.2%	-.044	-2.60	[.008]**	-0.4%
	Bad	.392	13.55	[.000]***	3.2%	.039	1.60	[.000]***	0.4%
	Very Bad	.491	11.13	[.000]***	4%	.209	5.71	[.000]***	2.1%
Education	Secondary finished	.034	1.66	[.095]*	0.2%	.055	3.07	[.045]**	0.5%
	Secondary not finished	.163	8.50	[.000]***	1.3%	.346	21.22	[.000]***	3.6%
Housing allowance		-.566	-26.37	[.000]***	-4.6%	-.244	-10.86	[.000]***	-2.5%
Number of Observations		130639				138130			
Log Likelihood		-20753.7				-27432.9			
Mean of Dependent Variable		.042198				.053993			
Percentage Correct Prediction		95.7%				94.5%			
Pseudo R^2		.033154				.023387			

***significant at 1% level; **significant at 5% level.

Table 5 Households lacking adequate heating facilities: results of a Probit model for 1994 and 1995

Dependent Variable: FACILITY		1994				1995			
Explanatory Variables		Coeff	t Stat	P-Value	Marginal Effects	Coeff	t Stat	P-Value	Marginal Effects
Constant		-1.73	-54.57	-	-	-1.73	-53.12	-	-
Age Group	16-34	.156	7.69	[.000]***	3.4%	.196	9.49	[.000]***	4%
	35-44	.080	3.99	[.000]***	1.8%	.056	2.75	[.006]**	1.1%
	45-54	-.014	-.76	[.445]	-0.3%	-.036	-1.84	[.066]*	-0.7%
	55-64	.058	3.42	[.001]**	1.2%	.546	.31	[.754]	0.1%
Household Composition	Number of Adults	.058	15.18	[.000]***	1.2%	.028	7.25	[.000]***	0.5%
	Number of Children	.015	2.86	[.004]**	0.3%	.021	3.81	[.000]***	0.4%
Marital Status	Separated	.136	3.55	[.000]***	3%	.134	3.30	[.001]**	2.7%
	Divorced	.046	1.76	[.078]*	1%	.059	2.17	[.029]**	1.2%
	Widowed	.048	2.68	[.007]**	1%	.055	3.05	[.002]**	1.1%
	Single	.122	8.96	[.000]***	2.7%	.100	7.25	[.000]***	2%
Income	Self employed	.213	15.27	[.000]***	4.7%	.197	13.97	[.000]***	4%
	Pension	.095	6.44	[.000]***	2.1%	.037	2.48	[.013]**	0.7%
	Unemployed	.028	1.00	[.317]	0.6%	-.495	-.15	[.878]	-0.1%
	Other benefit	.112	5.75	[.000]***	2.5%	.100	4.80	[.000]***	2%
	Private Income	.141	4.47	[.000]***	3.1%	.042	1.18	[.234]	0.8%
Tenure	Tenant	.411	36.87	[.000]***	9.1%	-.031	-2.70	[.007]**	-0.6%
	Rent free	.466	23.60	[.000]***	10%	.160	7.16	[.000]***	3.2%

Accommodation type	Semi-detached Terraced	-.192	-16.79	[.000]***	-4.3%	-.326	-.26	[.791]	0%
	Apartment Small	-.093	-7.22	[.000]***	-2%	.013	.96	[.334]	0.2%
	Apartment Large	-.421	-29.57	[.000]***	-9.4%	.767	.58	[.558]	0.1%
	Other Accommodation	.188	6.31	[.000]***	4.2%	-.091	-3.57	[.000]***	-1.8%
Health Status	Good	-.029	-2.55	[.011]**	-0.6%	-.518	-.43	[.667]	-0.1%
	Fair	.168	12.29	[.000]***	3.7%	.198	14.03	[.000]***	4%
	Bad	.442	24.31	[.000]***	9.8%	.503	26.78	[.000]***	10.3%
	Very Bad	.491	18.47	[.000]***	10.9%	.554	19.24	[.000]***	11.4%
Education	Secondary finished	.820	.567	[.571]	0.1%	.020	1.32	[.184]	0.4%
	Secondary not finished	.414	31.37	[.000]***	9.2%	.449	32.16	[.000]***	9.2%
Housing allowance		.056	2.86	[.004]**	1.2 %	.018	.87	[.383]	0.3 %
Number of Observations			128045				127912		
Log Likelihood			-51706.4				-47941.7		
Mean of Dependent Variable			.155766				.133506		
Percentage Correct Prediction			84.4%				86.6%		
Pseudo R^2			.057798				0.36574		

***significant at 1% level; **significant at 5% level.

Table 6 Households lacking adequate heating facilities: results of a Probit model for 1996 and 1997

Dependent Variable: FACILITY

Explanatory Variables		1996				1997			
		Coeff	t Stat	P-Value	Marginal Effects	Coeff	t Stat	P-Value	Marginal Effects
Constant		-1.74	-52.09	-	-	-2.14	-62.31	-	-
Age Group	16-34	.177	8.30	[.000]***	3.3%	.176	8.02	[.000]***	2.9%
	35-44	.057	2.73	[.006]**	1.1%	.085	3.87	[.000]***	1.4%
	45-54	-.016	-.84	[.400]	-0.3%	-.027	-1.31	[.190]	-0.4%
	55-64	-.028	-1.57	[.115]	-0.5%	.125	.06	[.946]	0%
Household Composition	Number of Adults	.024	6.10	[.000]***	0.4%	.077	18.59	[.000]***	1.3%
	Number of Children	.034	6.16	[.000]***	0.6%	.032	5.70	[.000]***	0.5%
Marital Status	Separated	.154	3.54	[.000]***	2.9%	.174	4.19	[.000]***	2.9%
	Divorced	.115	4.29	[.000]***	2.1%	.025	.94	[.345]	0.4%
	Widowed	.089	4.78	[.000]***	1.6%	.100	5.27	[.000]***	1.7%
	Single	.117	8.28	[.000]***	2.2%	.086	5.94	[.000]***	1.4%
Income	Self employed	.168	11.56	[.000]***	3.2%	.219	14.33	[.000]***	3.7%
	Pension	.031	2.07	[.038]**	0.6%	.141	8.90	[.000]***	2.3%
	Unemployed	.126	3.89	[.000]***	2.3%	.017	.51	[.608]	0.2%
	Other benefit	.076	3.60	[.000]***	1.4%	.043	2.01	[.044]**	0.7%
	Private Income	-.015	-.39	[.689]	-0.2%	.052	1.42	[.155]	0.8%
Tenure	Tenant	-.102	-8.61	[.000]***	-1.9%	.324	26.40	[.000]***	5.4%
	Rent free	.090	3.85	[.000]***	1.7%	.433	20.62	[.000]***	7.3%

Accommodation type	Semi-detached Terraced	-.092	-7.21	[.000]***	-1.7%	-.122	-9.96	[.000]***	-2%
	Apartment Small	-.035	-2.43	[.015]**	-0.6%	-.099	-7.07	[.000]***	-1.6%
	Apartment Large	.162	12.55	[.000]***	3%	-.432	-27.34	[.000]***	-7.3%
	Other Accommodation	-.408	-13.22	[.000]***	7.7%	-.248	-9.23	[.000]***	-4.2%
Health Status	Good	.035	2.80	[.005]**	0.6%	.056	4.42	[.000]***	0.9%
	Fair	.241	16.25	[.000]***	4.5%	.251	17.03	[.000]***	4.2%
	Bad	.557	28.84	[.000]***	10.6%	.522	27.17	[.000]***	8.8%
	Very Bad	.575	19.10	[.000]***	10.9%	.559	18.80	[.000]***	9.4%
Education	Secondary finished	-.035	-2.25	[.024]**	-0.6%	-.049	-3.21	[.000]**	-0.8%
	Secondary not finished	.411	29.27	[.000]***	7.8%	.485	35.94	[.000]***	8.1%
Housing allowance		-.415	-.19	[.843]	0%	.108	4.88	[.000]***	1.8%
Number of Observations		130639				138130			
Log Likelihood		-45534.8				-42951.1			
Mean of Dependent Variable		.120840				.106277			
Percentage Correct Prediction		87.9%				89.3%			
Pseudo R^2		.040669				.056184			

***significant at 1% level; **significant at 5% level.

Table 7 Presence of central heating: results of a Probit model for 1994 and 1995

Dependent Variable: CRLHEAT		1994				1995			
Explanatory Variables		Coeff	t Stat	P-Value	Marginal Effects	Coeff	t Stat	P-Value	Marginal Effects
Constant		2.32	77.78	-	-	2.04	68.11	-	-
Age Group	16-34	-.133	-7.46	[.000]***	-4.2%	-.145	-8.22	[.000]***	-4.5%
	35-44	-.886	-.50	[.616]	-0.2%	-.020	-1.14	[.252]	-0.6%
	45-54	.122	7.36	[.000]***	3.9%	.137	8.24	[.000]***	4.2%
	55-64	.067	4.53	[.000]***	2.1%	.075	5.05	[.000]***	2.3%
Household Composition	Number of Adults	-.109	-32.01	[.000]***	-3.4%	-.093	-27.83	[.000]***	-2.9%
	Number of Children	-.034	-7.13	[.000]***	-1.1%	-.865	-1.77	[.076]*	-0.2%
Marital Status	Separated	-.260	-7.51	[.000]***	-8.3%	-.293	-8.27	[.000]***	-9.1%
	Divorced	.086	3.52	[.000]***	2.7%	.035	1.46	[.144]	1.1%
	Widowed	-.086	-5.44	[.000]***	-2.7%	-.103	-6.54	[.000]***	-3.2%
	Single	-.150	-12.56	[.000]***	-4.8%	-.107	-8.96	[.000]***	-3.3%
Income	Self employed	-.134	-10.80	[.000]***	-4.3%	-.196	-15.96	[.000]***	-6.1%
	Pension	-.134	-10.39	[.000]***	-4.3%	-.156	-12.16	[.000]***	-4.8%
	Unemployed	-.372	-15.25	[.000]***	-11.9%	-.372	-13.97	[.000]***	-11.6%
	Other benefit	-.235	-13.00	[.000]***	-7.5%	-.168	-8.95	[.000]***	-5.2%
	Private Income	-.038	-1.37	[.169]	-1.2%	-.090	-3.01	[.003]**	-2.8%
Tenure	Tenant	-.078	-7.90	[.000]***	-2.5%	-.168	-17.25	[.000]***	-5.2%
	Rent free	-.499	-26.70	[.000]***	-16%	.037	1.84	[.065]*	1.1%

Accommodation type									
	Semi-detached	.011	1.15	[.246]	-0.3%	.328	30.98	[.000]***	10.2%
	Terraced								
	Apartment Small	-.091	-8.00	[.000]***	-2.9%	.381	31.20	[.000]***	11.9%
	Apartment Large	.026	2.26	[.024]**	0.8%	.165	14.92	[.000]***	5.1%
	Other Accommodation	-.019	-.667	[.504]	-0.6%	.135	6.49	[.000]***	4.2%
Health Status	Good	-.197	-19.58	[.000]***	-6.3%	-.151	-14.91	[.000]***	-4.7%
	Fair	-.282	-23.52	[.000]***	-9%	-.246	-20.29	[.000]***	-7.6%
	Bad	-.695	-41.64	[.000]***	-22.2%	-.629	-37.09	[.000]***	-19.6%
	Very Bad	-.564	-22.39	[.000]***	-18%	-.493	-18.22	[.000]***	-15.4%
Education	Secondary finished	-.076	-6.21	[.000]***	-2.4%	-.053	-4.18	[.000]***	-1.6%
	Secondary not finished	-.805	-70.94	[.000]***	-25.8%	-.774	-66.11	[.000]***	-24.1%
Housing allowance		-.745	-37.27	[.000]***	-23.8 %	-.678	-32.61	[.000]***	-21.1 %
Number of Observations		128045				127912			
Log Likelihood		-54516.4				-70449.8			
Mean of Dependent Variable		.167941				.686042			
Percentage Correct Prediction		83.2%				71.5%			
Pseudo R^2		.053848				.140859			

***significant at 1% level; **significant at 5% level.

Table 8 Presence of central heating: results of a Probit model for 1996 and 1997

Dependent Variable: CRLHEAT		1996				1997			
Explanatory Variables		Coeff	t Stat	P-Value	Marginal Effects	Coeff	t Stat	P-Value	Marginal Effects
Constant		2.06	67.42	-	-	.760	30.10	-	-
Age Group	16-34	-.126	-7.05	[.000]***	-3.7%	-.564	-.33	[.739]	-0.1%
	35-44	.283	-.15	[.874]	0%	.070	4.17	[.000]***	2.4%
	45-54	.150	8.95	[.000]***	4.4%	.100	6.34	[.000]***	3.4%
	55-64	.107	7.16	[.000]***	3.2%	.074	5.20	[.000]***	2.5%
Household Composition	Number of Adults	-.093	-27.57	[.000]***	-2.7%	-.028	-8.51	[.000]***	-0.9%
	Number of Children	-.015	-3.02	[.002]**	-0.4%	.013	2.97	[.003]**	0.4%
Marital Status	Separated	-.271	-7.28	[.000]***	-8.1%	-.073	-2.20	[.027]**	-2.5%
	Divorced	.025	1.06	[.288]	0.7%	.076	3.84	[.000]***	2.6%
	Widowed	-.127	-7.99	[.000]***	-3.8%	.013	.91	[.361]	0.4%
	Single	-.075	-6.30	[.000]***	-2.2%	.041	3.76	[.000]***	1.4%
Income	Self employed	-.154	-12.38	[.000]***	-4.6%	.048	3.89	[.000]***	1.6%
	Pension	-.127	-9.91	[.000]***	-3.8%	-.140	-11.31	[.000]***	-4.8%
	Unemployed	-.399	-14.48	[.000]***	-11.9%	-.241	-9.70	[.000]***	-8.2%
	Other benefit	-.154	-8.20	[.000]***	-4.5%	-.039	-2.34	[.019]**	-1.3%
	Private Income	-.020	-.63	[.528]	-0.6%	.220	7.65	[.000]***	7.5%
Tenure	Tenant	.017	1.71	[.087]*	0.5%	-.082	-8.70	[.000]***	-2.8%
	Rent free	-.076	-3.77	[.000]***	-2.2%	-.304	-16.86	[.000]***	-10.4%

Accommodation type	Semi-detached Terraced	.358	33.62	[.000]***	10.7%	.160	16.99	[.000]***	5.5%
	Apartment Small	.401	32.57	[.000]***	11.9%	.143	12.88	[.000]***	4.9%
	Apartment Large	.083	7.47	[.000]***	2.4%	.123	11.46	[.000]***	4.2%
	Other Accommodation	.442	19.75	[.000]***	13.1%	.027	1.43	[.150]	0.9%
Health Status	Good	-.161	-15.57	[.000]***	-4.8%	.209	23.14	[.000]***	7.1%
	Fair	-.238	-19.23	[.000]***	-7.1%	.169	15.55	[.000]***	5.7%
	Bad	-.617	-35.98	[.000]***	-18.4%	-.107	-6.94	[.000]***	-3.6%
	Very Bad	-.539	-19.46	[.000]***	-16%	-.045	-1.77	[.077]*	-1.5%
Education	Secondary finished	-.070	-5.45	[.000]***	-2.1%	-.117	-11.17	[.000]***	-4%
	Secondary not finished	-.780	-65.85	[.000]***	-23.2%	-.602	-61.11	[.000]***	-20.6%
Housing allowance		-.680	-32.17	[.000]***	-20.3 %	-.115	-7.00	[.000]***	-3.9 %
Number of Observations		130639				138130			
Log Likelihood		-68958.1				-83135.2			
Mean of Dependent Variable		.713359				.668892			
Percentage Correct Prediction		73.5%				67.8%			
Pseudo R^2		.141218				.065727			

***significant at 1% level; **significant at 5% level.

Table 9 Presence of damp: results of a Probit model for 1994 and 1995

Dependent Variable: DAMP		1994				1995			
Explanatory Variables		Coeff	t Stat	P-Value	Marginal Effects	Coeff	t Stat	P-Value	Marginal Effects
Constant		-1.689	-55.03	-	-	-1.482	-47.52	-	-
Age Group	16-34	.185	9.25	[.000]***	4.3%	.172	8.50	[.000]***	3.7%
	35-44	.068	3.40	[.001]**	1.6%	.048	2.40	[.016]**	1%
	45-54	.016	.84	[.396]	0.3%	-.017	-.92	[.356]	-0.3%
	55-64	.031	1.86	[.062]*	0.7%	-.109	-.62	[.995]	0%
Household Composition	Number of Adults	.065	17.36	[.000]***	1.5%	.034	9.10	[.000]***	0.7%
	Number of Children	.055	10.69	[.000]***	1.3%	.062	11.92	[.000]***	1.3%
Marital Status	Separated	.192	5.20	[.000]***	4.5%	.206	5.31	[.000]***	4.5%
	Divorced	.127	5.11	[.000]***	3%	.212	8.52	[.000]***	4.6%
	Widowed	.026	1.49	[.135]	0.6%	.079	4.39	[.000]***	1.7%
	Single	.117	8.94	[.000]***	2.7%	.111	8.25	[.000]***	2.4%
Income	Self employed	.014	1.03	[.301]	0.3%	-.011	-.81	[.413]	-0.2%
	Pension	.044	3.03	[.002]**	1%	.133	.08	[.929]	0%
	Unemployed	.145	5.52	[.000]***	3.4%	.164	5.61	[.000]***	3.6%
	Other benefit	.068	3.57	[.000]***	1.6%	.125	6.29	[.000]***	2.7%
	Private Income	.018	.58	[.560]	0.4%	.039	1.15	[.248]	0.8%
Tenure	Tenant	.453	41.82	[.000]***	10.7%	.094	8.52	[.000]***	2%
	Rent free	.503	25.84	[.000]***	11.8%	.030	1.30	[.191]	0.6%

Category	Subcategory	Coef	t	p-value	%	Coef	t	p-value	%
Accommodation type	Semi-detached	.044	4.08	[.000]***	1%	-.132	-11.04	[.000]***	-2.8%
	Terraced								
	Apartment Small	-.212	-16.05	[.000]***	-5%	-.154	-11.15	[.000]***	-3.3%
	Apartment Large	-.366	-26.49	[.000]***	-8.6%	-.164	-12.64	[.000]***	-3.5%
	Other Accommodation	-.098	-3.05	[.002]**	-2.3%	-.055	-2.34	[.019]**	-1.2%
Health Status	Good	.108	9.47	[.000]***	2.5%	.130	11.01	[.000]***	2.8%
	Fair	.280	20.78	[.179]	6.6%	.312	22.42	[.000]***	6.8%
	Bad	.556	30.7	[.000]***	13.1%	.577	30.81	[.000]***	12.6%
	Very Bad	.606	22.89	[.000]***	14.3%	.635	22.14	[.000]***	13.8%
Education	Secondary finished	-.016	-1.24	[.213]	-0.4%	-.041	-2.93	[.003]**	-0.9%
	Secondary not finished	.280	22.13	[.000]***	6.6%	.250	19.06	[.000]***	5.4%
Housing allowance		-.027	-1.46	[.143]	-0.6 %	-.108	-5.61	[.000]***	-2.3%
Number of Observations		128045				127912			
Log Likelihood		-54516.4				-50777.9			
Mean of Dependent Variable		.167941				.143614			
Percentage Correct Prediction		83.2%				85.6%			
Pseudo R^2		.053848				.029082			

***significant at 1% level; **significant at 5% level.

Table 10 Presence of damp: results of a Probit model for 1996 and 1997

Dependent Variable: DAMP		1996				1997			
Explanatory Variables		Coeff	t Stat	P-Value	Marginal Effects	Coeff	t Stat	P-Value	Marginal Effects
Constant		-1.67	-51.99	-	-	-2.25	-66.84	-	-
Age Group	16-34	.166	7.93	[.000]***	3.3%	.133	6.10	[.000]***	2.3%
	35-44	.067	3.24	[.001]**	1.3%	.100	4.62	[.000]**	1.7%
	45-54	-.016	-.81	[.413]	-0.3%	.010	.496	[.620]	0.1%
	55-64	-.017	-.95	[.339]	-0.3%	.897	.481	[.630]	0.1%
Household Composition	Number of Adults	.051	13.34	[.000]***	1%	.107	26.31	[.000]***	1.8%
	Number of Children	.071	13.29	[.000]***	1.4%	.059	10.74	[.000]***	1%
Marital Status	Separated	.179	4.25	[.000]***	3.6%	.259	6.51	[.000]***	4.5%
	Divorced	.138	5.42	[.000]***	2.8%	.073	2.81	[.005]**	1.2%
	Widowed	.100	5.43	[.000]***	2%	.156	8.24	[.000]***	2.7%
	Single	.112	8.14	[.000]***	2.2%	.146	10.29	[.000]***	2.5%
Income	Self employed	.120	.08	[.936]	0%	.045	2.86	[.004]**	0.8%
	Pension	.035	2.34	[.019]**	0.7%	.125	7.91	[.000]***	2.2%
	Unemployed	.302	10.16	[.000]***	6.1%	.184	6.03	[.000]***	3.2%
	Other benefit	.163	8.16	[.000]***	3.3%	.138	6.79	[.000]***	2.4%
	Private Income	-.023	-.615	[.538]	-0.4%	.132	3.74	[.000]***	2.3%
Tenure	Tenant	.050	4.39	[.000]***	1%	.358	29.82	[.000]***	6.3%
	Rent free	.065	2.77	[.005]**	1.3%	.465	22.36	[.000]***	8.1%

Category	Subcategory	Coef	t	p	%	Coef	t	p	%
Accommodation type	Semi-detached	-.133	-10.91	[.000]***	-2.6%	.115	9.95	[.000]***	2%
	Terraced								
	Apartment Small	-.215	-14.94	[.000]***	-4.3%	-.221	-15.00	[.000]***	-3.8%
	Apartment Large	-.068	-5.27	[.000]***	-1.3%	-.382	-24.85	[.000]***	-6.7%
	Other Accommodation	-.254	-9.73	[.000]***	-5.1%	-.356	-12.69	[.000]***	-6.2%
Health Status	Good	.101	8.32	[.000]***	2%	.180	14.17	[.000]***	3.1%
	Fair	.274	19.11	[.000]***	5.5%	.312	21.15	[.000]***	5.5%
	Bad	.571	29.83	[.000]***	11.5%	.608	31.65	[.000]***	10.7%
	Very Bad	.654	22.14	[.000]***	13.2%	.643	21.70	[.000]***	11.3%
Education	Secondary finished	-.319	-.21	[.828]	0%	-.051	-3.52	[.000]***	-0.8%
	Secondary not finished	.329	24.32	[.000]***	6.6%	.345	26.63	[.000]***	6%
Housing allowance		-.065	-3.31	[.001]**	-1.3%	.098	4.60	[.000]***	1.7%
Number of Observations		130639				138130			
Log Likelihood		-48208.4				-44812.9			
Mean of Dependent Variable		.129747				.111909			
Percentage Correct Prediction		87%				88.8%			
Pseudo R^2		.034251				.055268			

***significant at 1% level; **significant at 5% level.

Table 11 Presence of rot: results of a Probit model for 1994 and 1995

Dependent Variable: ROT				1994			1995		
Explanatory Variables		Coeff	t Stat	P-Value	Marginal Effects	Coeff	t Stat	P-Value	Marginal Effects
Constant		-1.88	-54.56	-	-	-1.66	-48.03	-	-
Age Group	16-34	.071	3.18	[.001]**	1.2%	.130	5.74	[.000]***	2.1%
	35-44	.015	.708	[.479]	0.2%	.031	1.40	[.160]	0.5%
	45-54	-.365	-.01	[.986]	0%	.679	.31	[.753]	0.1%
	55-64	.030	1.57	[.116]	0.5%	.010	.55	[.577]	0.1%
Household Composition	Number of Adults	.025	5.90	[.000]***	0.4%	.146	.34	[.731]	0%
	Number of Children	.058	10.19	[.000]***	1%	.059	10.19	[.000]***	0.9%
Marital Status	Separated	.214	5.32	[.000]***	3.7%	.247	5.90	[.000]***	4%
	Divorced	.191	7.15	[.000]***	3.3%	.224	8.28	[.000]***	3.6%
	Widowed	.067	3.39	[.001]**	1.1%	.121	6.15	[.000]***	2%
	Single	.193	13.17	[.000]***	3.3%	.172	11.58	[.000]***	2.8%
Income	Self employed	-.725	-.43	[.663]	-0.1%	-.057	-3.45	[.001]**	-0.9%
	Pension	.012	.76	[.441]	0.2%	-.021	-1.26	[.205]	-0.3%
	Unemployed	.100	3.44	[.001]**	1.7%	.073	2.20	[.027]**	1.2%
	Other benefit	.099	4.78	[.000]***	1.7%	.128	5.91	[.000]***	2.1%
	Private Income	-.209	-5.17	[.000]***	-3.6%	-.164	-3.93	[.000]***	-2.7%
Tenure	Tenant	.546	45.55	[.000]***	9.4%	.034	2.76	[.006]**	0.5%
	Rent free	.460	21.41	[.000]***	7.9%	-.038	-1.44	[.149]	-0.6%

Category	Subcategory	Coef.	z	[p-value]	%	Coef.	z	[p-value]	%
Accommodation type	Semi-detached Terraced	.019	1.54	[.122]	0.3%	-.071	-5.31	[.000]***	-1.1%
	Apartment Small	-.205	-13.71	[.000]***	-3.5%	-.038	-2.54	[.011]**	-0.6%
	Apartment Large	-.378	-23.59	[.000]***	-6.5%	-.040	-2.81	[.005]**	-0.6%
	Other Accommodation	.137	4.19	[.000]***	2.3%	-.874	-.33	[.741]	-0.1%
Health Status	Good	.056	4.28	[.000]***	0.9%	.138	10.26	[.000]***	2.2%
	Fair	.244	16.07	[.000]***	4.2%	.310	19.76	[.000]***	5.1%
	Bad	.512	25.63	[.000]***	8.8%	.591	28.74	[.000]***	9.7%
	Very Bad	.523	17.97	[.000]***	9%	.634	20.49	[.000]***	10.4%
Education	Secondary finished	.628	.405	[.685]	0.1%	-.035	-2.19	[.028]**	-0.5%
	Secondary not finished	.220	15.29	[.000]***	3.8%	.235	15.94	[.000]***	3.8%
Housing allowance		.061	2.96	[.003]**	1%	-.109	-5.21	[.000]***	-1.8%
Number of Observations		128045				127912			
Log Likelihood		-40874.9				-38996.7			
Mean of Dependent Variable		.107095				.095753			
Percentage Correct Prediction		89.2%				90.4%			
Pseudo R^2		.042783				.021697			

***significant at 1% level; **significant at 5% level.

Table 12 Presence of rot: results of a Probit model for 1996 and 1997

Dependent Variable: ROT		1996				1997			
Explanatory Variables		Coeff	t Stat	P-Value	Marginal Effects	Coeff	t Stat	P-Value	Marginal Effects
Constant		-1.76	-48.36	-	-	-2.27	-59.69	-	-
Age Group	16-34	.124	5.19	[.000]***	1.8%	.496	.01	[.984]	0%
	35-44	.050	2.12	[.033]**	0.7%	.033	1.34	[.179]	0.4%
	45-54	.043	-1.93	[.053]*	0.6%	-.016	-.69	[.489]	-0.2%
	55-64	-.206	-.10	[.920]	0%	-.798	-.37	[.709]	0%
Household Composition	Number of Adults	-0.12	-2.82	[.005]**	-0.1%	.053	11.24	[.000]***	0.6%
	Number of Children	.062	10.41	[.000]***	0.9%	.069	11.07	[.000]***	0.8%
Marital Status	Separated	.259	5.75	[.000]***	3.7%	.313	7.28	[.000]***	3.9%
	Divorced	.208	7.58	[.000]***	3%	.162	5.84	[.000]***	2%
	Widowed	.132	6.39	[.000]***	1.9%	.155	7.29	[.000]***	1.9%
	Single	.197	12.81	[.000]***	2.8%	.248	15.41	[.000]***	3%
Income	Self employed	-.084	-4.71	[.000]***	-1.2%	-.013	-.712	[.476]	-0.1%
	Pension	-.026	-1.55	[.121]	-0.3%	.066	3.63	[.000]***	0.8%
	Unemployed	.222	6.68	[.000]***	3.2%	.097	2.81	[.005]**	1.2%
	Other benefit	.150	6.82	[.000]***	2.1%	.098	4.35	[.000]***	1.2%
	Private Income	-.088	-2.02	[.043]**	-1.2%	-.041	-.95	[.341]	-0.5%
Tenure	Tenant	.016	1.25	[.210]	0.2%	.476	35.44	[.000]***	5.9%
	Rent free	-.043	-1.53	[.125]	-0.6%	.515	22.54	[.000]***	6.4%

Accommodation type	Semi-detached Terraced	-.094	-6.74	[.000]***	-1.3%	.108	8.20	[.000]***	1.3%
	Apartment Small	-.138	-8.44	[.000]***	-2%	-.233	-13.67	[.000]***	-2.9%
	Apartment Large	.018	1.26	[.206]	0.2%	-.417	-22.82	[.000]***	-5.2%
	Other Accommodation	-.228	-7.46	[.000]***	-3.3%	-.291	-9.36	[.000]***	-3.6%
Health Status	Good	.118	8.37	[.000]***	1.7%	.198	13.16	[.000]***	2.4%
	Fair	.276	16.77	[.000]***	4%	.336	19.61	[.000]***	4.2%
	Bad	.548	25.42	[.000]***	7.9%	.589	26.90	[.000]***	7.3%
	Very Bad	.632	19.40	[.000]***	9.1%	.677	20.84	[.000]***	8.4%
Education	Secondary finished	-.016	-.99	[.318]	-0.2%	-.091	-5.46	[.000]**	-1.1%
	Secondary not finished	.260	16.95	[.000]***	3.7%	.257	17.48	[.000]***	3.2%
Housing allowance		-.060	-2.75	[.006]**	-0.8 %	.050	2.17	[.029]***	0.6%
Number of Observations				130639				138130	
Log Likelihood				-35432.2				-32360.9	
Mean of Dependent Variable				.081144				.070318	
Percentage Correct Prediction				91.8%				92.9%	
Pseudo R^2				.021062				.041182	

***significant at 1% level; **significant at 5% level.

APPENDIX II

Table 1 Ability to heat home to an adequate level: results of an ordered-Probit model

Dependent Variable: HEATHOME		Statistics				Marginal Effects			
Explanatory Variables		Estimate	Std. Error	t-stat	P-value	Yes, without any problems	Usually with occasional difficulty	Usually not,	No, not at all
Constant		-.741555	.14000	-5.2967	[.000]***				
Age Group	35-45	.042964	.13380	.321104	[.748]	-1%	0%	0%	0%
	46-55	.157665	.12787	1.23295	[.218]	-3.7%	2.3%	0%	0%
	56-65	.201874	.14175	1.42407	[.154]	-4.8%	2.9%	0%	1%
	Over 65	.230216	.14561	1.58102	[.114]	-5.5%	3.4%	0%	1.1%
Rural		.194284	.08175	.023764	[.981]	0%	0%	0%	0%
Education	Upper Secondary Technical	.090806	.12360	.734644	[.463]	-2.1%	1.3%	0%	0%
	Upper Secondary Leav. Cert	-.046102	.0992	-.46447	[.642]	1.1%	0%	0%	0%
	Degree	-.326337	.13487	-2.4196	[.016]**	7.8%	-4.8%	-1.3%	-1.6%
Tenure	Own with a mortgage	-.017389	.10763	-.16155	[.872]	0%	0%	0%	0%
	Tenant	.361182	.1644	2.19605	[.028]**	-8.6%	5.3%	1.4%	1.8%

	Rent free	.446028	.11753	3.79479	[.000]***	-10.6%	6.6%	1.8%	2.2%
Year Built	41-79	-.109584	.09933	-1.1031	[.270]	2.6%	-1.6%	0%	0%
	80-89	-.286160	.12787	-2.2378	[.025]**	6.8%	-4.2%	-1.1%	-1.4%
	90-2002	-.223010	.14396	-1.5490	[.121]	5.3%	-3.3%	-1%	-1.1%
Marital status	Couple	-.386882	.09970	-3.8804	[.000]***	9.2%	-5.7%	-1.5%	-1.9%
	Separated /Divorced/ Widowed	-.073634	.14139	-.52075	[.603]	1.7%	-1%	0%	0%
Household Composition	Kids	.921553	.75377	1.22259	[.221]	0%	0%	0%	0%
	Adults	-.011089	.78523	-1.4121	[.158]	0%	0%	0%	0%
Number of Observations					1500				
Log Likelihood					-845.318				
Mean of Dependent Variable					1.24133				
Pseudo R^2					.058863				

Bibliography

Abel-Smith, B. and Townsend, P. (1965). *The Poor and the Poorest: A New Analysis of the Ministry of Labour's Family Expenditure Surveys of 1953-54 and 1960*, G. Bell & Sons, London.

Alderson, M.R. (1985). 'Season and Mortality', *Health Trends*, **17**, pp. 87-96.

Allamby, L. (1995). *Paying the Price: The Changing Face of Fuel Poverty*, Neighbourhood Energy Action and Heat & Energy Action Tallaght, Dublin.

Anderson, T.W. and Le Riche, W.H. (1970). 'Cold Weather and Myocardial Infarction', *Lancet*, i, pp. 291-6.

Andreb, H.J. (ed.) (1998). *Empirical Poverty Research in a Comparative Perspective*, Ashgate, Aldershot.

Atkinson, A.B. (1970). 'On the Measurement of Inequality', *Journal of Economic Theory*, **2**, pp. 244-263.

Atkinson, A.B. (1987). 'On the Measurement of Poverty', *Econometrica*, **55**(4), pp. 749-64.

Baker-Blocker, A. (1982). 'Winter Deaths and Cardiovascular Mortality in Minneapolis-St. Paul', *American Journal of Public Health*, **72**, pp. 88-97.

Barresi, C.M., Ferraro, K.F. and Hobet, L.L. (1984). 'Environmental Satisfaction, Sociability, and Well-Being among Urban Elderly', *International Journal of Aging & Human Development*, **18**(4), pp. 277-93.

Backlund, E., Sorlie, P.D. and Johnson, N.J. (1999). 'A Comparison of the Relationships between Education and Income with Mortality: the National Longitudinal Mortality Study', *Social Science Medicine*, **49**, pp. 1373-84.

Becker, S. and Weng, S. (1998). 'Seasonal Patterns of Deaths in Matlab, Bangladesh', *International Journal of Epidemiology*, **27**, pp. 814-23.

Bellido, N.P., Jano, M.D., López Ortega, F.J., Martin-Guzman, M.P. and Toledo, M.I. (1998). 'The Measurement and Analysis of Poverty and Inequality: an Application to Spanish Conurbations', *International Statistical Review*, **66**(1), pp. 115-31.

Berthound, R. (1976). *The Disadvantages of Inequality: A Study of Social Deprivation*, MacDonald and James, London.

Beveridge, W. (1942). *Social Insurance and Applied Services*, HM Stationary Office, London.

Blackburn, M.L. (1998). 'The Sensitivity of International Poverty Comparisons', *Review of Income and Wealth*, **44**(4), pp. 449-72.

Blasnik, M. (1998). *Impact Evaluation of Ohio's Home Weatherization Assistance Program: 1994 Program Year*, Proctor Engineering Group, Ohio.

Boardman, B. (1986). 'Seasonal Mortality and Cold Homes', Proceedings of the University of Warwick's *Unhealthy Housing: A Diagnosis* conference, 14th-16th December, 1986.

Boardman, B. (1991). *Fuel Poverty: From Cold Homes to Affordable Warmth*, Belhaven Press, London.

Boardman, B. (1993). 'Opportunities and Constraints Posed by Fuel Poverty on Policies to Reduce the Greenhouse Effect in Britain', *Applied Energy*, **44**, pp. 185-95.

Boardman, B. (1998). *Energy, Efficiency and Equity*, Oxford Environment Lecture Series, University of Oxford.

Boardman, B. (2000). 'Introduction', in J. Rudge and F. Nicol (eds.) (2000), *Cutting the Cost of Cold: Affordable Warmth for Healthier Homes*, E&FN Spon, London.

Boyle, P. and Maisonneuve, P. (1995). 'Lung Cancer and Tobacco Smoking', *Lung Cancer*, **12**(3), pp. 167-81.

Bradbury, B. and Jantti, M. (1999). *Child Poverty across Industrialized Nations*, Innocenti Occasional Papers, Economic and Social Policy Series 71, UNICEF, Florence.

Bradshaw, J. (1980). 'Cold Conditions – the Social Circumstances', Paper presented at the *Cold Conditions* conference, January 1980, National Fuel Poverty Forum, London.

Bradshaw, J. and Millar, J. (1991). *Lone Parents in the UK*, HM Stationary Office, London.

Bradshaw, J. (1999). 'Child Poverty in a Comparative Perspective', *European Journal of Social Security*, **1**(4), pp. 383-404.

BRE (Building Research Establishment) (1995). *Domestic Energy Fact File*, Building Research Establishment, Watford.

Brainerd, E. (1998). 'Market Reform and Mortality in Transitional Economies', *World Development*, **26**(11), pp. 2013-27.

Brechling, V. and Smith, S. (1992). *The Pattern of Energy Efficiency Measures amongst Domestic Households in the UK* (Commentary No. 31), Institute for Fiscal Studies, London.

Brechling, V. and Smith, S. (1994). 'Household Energy Efficiency in the UK', *Fiscal Studies*, **15**(2), pp. 44-56.

Broadway, M.J. and Jesty, G. (1998). 'Are Canadian Inner Cities Becoming More Dissimilar? An Analysis of Urban Deprivation Indicators', *Urban Studies*, **35**(9), pp. 1,423-38.

Brophy, V., Clinch, J.P., Convery, F.J., Healy, J.D., King, C. and Lewis, J.O. (1999). *Homes for the 21st Century: the Costs and Benefits of Comfortable Housing*, Report for Energy Action by the Energy Research Group and Environmental Institute, University College, Dublin.

Browne, J. (1998). *The Relationship between Indoor Humidity, Fuel Poverty and Housing Conditions on Exacerbation, Symptoms and Lung function of Patients with Moderate and Severe Chronic Obstructive Pulmonary Disease*, Eaga Charitable Trust, London.

Brown, M. and Madge, N. (1982). *Despite the Welfare State*, Heinemann Educational Books, London.

Buhmann, B., Rainwater, L., Schmaus, G. and Smeeding, T. (1988). 'Equivalence Scales, Well-Being, Inequality and Poverty: Sensitivity Estimates across Ten Countries Using the Luxembourg Income Study Database', *Review of Income and Wealth*, **33**(2), pp. 115-42.

Burr, M.L., St Leger, A.S. and Yarnell, J.W.G. (1981). 'Wheezing, Dampness and Coal Fires', *Community Medicine*, **3**, pp. 203-9.

Callan, T., Nolan, B., Whelan, B.J., Hannan, D.F. and Creighton, S. (1989). *Poverty, Income and Welfare in Ireland*, Economic and Social Research Institute, Dublin.

Callan, T., Nolan, B. and Whelan, C.T. (1993). 'Resources, Deprivation and the Measurement of Poverty', *Journal of Social Policy*, **22**(2), pp. 141-72.

Callan, T., Nolan, B., Whelan, B.J., Whelan, C.T. and Williams, J. (1996). *Poverty in the 1990s: Evidence from the 1994 Living in Ireland Survey*, Oak Tree Press, Dublin.

Callan, T. and Nolan, B. (1999). *Tax and Welfare Changes, Poverty and Work Incentives in Ireland, 1987-1994*. Policy Research Series #34, Economic and Social Research Institute, Dublin.

Campbell, R. (1993). 'Fuel Poverty and Government Response', *Social Policy & Administration*, **27**(1), pp. 58-70.

Cantillon, S. (1997). Women and Poverty: Differences in Living Standards within Households, in A. Byrne and M. Leonard (eds.), *Women and Irish Society*, Beyond the Pale Publications, Belfast.

Cantillon, S. and Nolan, B. (1998). 'Are Married Women More Deprived than their Husbands?', *Journal of Social Policy*, **27**(2), pp. 151-71.

Caputo, R.K. (1995). 'Income Inequality and Family Poverty', *Families in Society*, **9**, pp. 604-15.

Carlsmith, R., Chandler, W., McMahon, J. and Santino, D. (1990). *Energy Efficiency: How Far Can We Go?*, Oak Ridge National Laboratory, Oak Ridge.

Chan, P.K.S., Sung, R.Y.T., Fung, K.S.C., Hui, M., Chik, K.W., Adeyemi-Doro, F.A.B. and Cheng, A.F. (1999). 'Epidemiology of Respiratory Syncytial Virus Infection among Paediatric Patients in Hong Kong: Seasonality and Disease Impact', *Epidemiology and Infection*, **123**(2), pp. 257-62.

Chapman, T. (1996). 'Domestic Energy Conservation in a Cold Climate', *Policy Studies*, **17**(4), pp. 299-313.

Charles, N. and Kerr, M. (1987). 'Just the Way It Is: Gender and Age Differences in Family Food Consumption', in J. Brannen and G. Wilson (eds.), *Give and Take in Families*, Allen and Unwin, London.

Clinch, J.P. (2002). 'Reconciling Rapid Economic Growth and Environmental Sustainability in Ireland', *Journal of the Statistical and Social Inquiry Society of Ireland*, XXX, pp. 159-218.

Clinch, J.P. and Healy, J.D. (1999a). 'Alleviating Fuel Poverty in Ireland: a Program for the 21st Century', *International Journal for Housing Science*, **23**(4), pp. 203-15.

Clinch, J.P. and Healy, J.D. (1999b). 'Housing Standards and Excess Winter Mortality in Ireland', *Environmental Studies Research Series* Working Paper 99/02, Department of Environmental Studies, University College, Dublin.

Clinch, J.P. and Healy, J.D. (2000a). 'Domestic Energy Efficiency in Ireland: Correcting Market Failure', *Energy Policy*, **28**(1), pp. 1-8.

Clinch, J.P. and Healy, J.D. (2000b). 'The Benefits of Residential Energy Conservation in the Light of the Luxembourg Agreement and the Gothenburg Protocol', *European Environment*, **10**, pp. 131-39.

Clinch, J.P. and Healy, J.D. (2000c). 'Housing Standards and Excess Winter Mortality', *Journal of Epidemiology & Community Health*, **54**(9), pp. 719-20.

Clinch, J.P. and Healy, J.D. (2000d). 'Evaluating the Health Benefits of Improving Domestic Energy Efficiency', *Environmental Studies Research Series* Working Paper 00/01, Department of Environmental Studies, University College, Dublin.

Clinch, J.P. and Healy, J.D. (2001). 'Cost-Benefit Analysis of Domestic Energy Efficiency', *Energy Policy*, **29**(2), pp. 113-24.

Clinch, J.P., Healy, J.D. and King, C. (2001). 'Modelling Improvements in Domestic Energy Efficiency', *Environmental Modelling & Software*, **16**, pp. 87-106.

Clinch, J.P. and Healy, J.D. (2003). 'Valuing Improvements in Comfort from Domestic Energy-Efficiency Retrofits Using a Trade-off Simulation Model', *Energy Economics*, **25**, pp. 565-583.

Cloke, P., Goodwin, M., Milbourne, P. and Thomas, C. (1995). 'Deprivation, Poverty and Marginalization in Rural Lifestyles in England and Wales', *Journal of Rural Studies*, **11**(4), pp. 351-65.

Collins, K.J. (1986). 'Low Indoor Temperatures and Morbidity in the Elderly', *Age and Ageing*, **15**, pp. 212-20.

Collins, K.J. (2000). 'Cold, Cold Housing and Respiratory Illness', in J. Rudge and F. Nicol (eds.), *Cutting the Cost of Cold: Affordable Warmth for Healthier Homes*, E&FN Spon, London.

Collins, K.J., Doré, C., Exton-Smith, A.N., Fox, R.H., Macdonald, I.C. and Woodward, P.M. (1977). 'Accidental Hypothermia and Impaired Temperature Homeostasis in the Elderly', *British Medical Journal*, **278**, pp. 353-6.

Collins, K.J., Exton-Smith, A.N. and Doré, C. (1981). 'Urban Hypothermia: Preferred Temperature and Thermal Perception on Old Age', *British Medical Journal*, **282**, pp. 175-177.

Collins, K.J., Exton-Smith, A.N. (1983). 'Thermal Homeostasis in Old Age', *Journal of the American Geriatric Society*, **31**, pp. 519-24.

Convery, F.J. (1978). *Applications of Economics in National Forest Planning*, U.S. Department of Agriculture and Duke University, Duke.

Convery, F.J. (1998). *A Guide to Policies for Energy Conservation: The European Experience*, Edward Elgar, Cheltenham.

Cornia, A.G. (1990). 'Child Poverty and Deprivation in Industrialized Countries: Recent Trends and Policy Options', UNICEF International Child Development Centre *Occasional papers* #2, Florence.

Coyle, C. (1996). 'Local and Regional Administrative Structures and Rural Poverty', in C. Curtin, T. Haase and H. Tovey (eds.), *Poverty in Rural Ireland*, Oak Tree Press, Dublin.

Cumper, G.E. (1981). 'National Incomes and Health: Implications of Some Recent Additions to the Data, *Journal of Tropical Medicine and Hygiene*, **84**(2), pp. 49-61.

Curwen. M. (1991). 'Excess Winter Mortality: a British Phenomenon?', *Health Trends*, **22**, pp. 169-75.

Dales, R.E., Burnett, R. and Zwanenburg, H. (1991). 'Adverse Health Effects among Adults Exposed to Home Dampness and Molds', *American Review of Respiratory Disease*, **142**, pp. 505-09.

Davies, H. and Joshi, H. (1994). 'Sex, Sharing and the Distribution of Income', *Journal of Social Policy*, **23**(3), pp. 301-40.

Davies, R. and Smith, W. (1998). *The Basic Necessities Survey: The Experience of Action Aid Vietnam*, Action Aid, London.

Deaton, A. (2001a). *Relative Deprivation, Inequality and Mortality*, NBER Working Paper, Center for Health and Well-Being, Princeton University.

Deaton, A. (2001b). *Health, Inequality and Economic Development*, NBER Working Paper, Center for Health and Well-Being, Princeton University.

DEFRA and DTI (Department of the Environment, Food and Rural Affairs and the Department of Trade and Industry) (2001). *The UK Fuel Poverty Strategy*, Her Majesty's Stationary Office, London.

Delhausse, B., Luttgens, A. and Perelman, S. (1993). 'Comparing Measures of Poverty and Relative Deprivation: an Example for Belgium', *Journal of Population Economics*, **6**, pp. 83-102.

DETR (Department of Environment, Transport and the Regions) (1999). *Fuel Poverty: A Programme for Warmer, Healthier Homes*, Her Majesty's Stationary Office, London.

Donaldson, G. C., Tchernjavskii, V. E., Ermakov, S. P., Bucher, K. and Keatinge, W. R. (1998). 'Winter Mortality and Cold Stress in Yekaterinburg, Russia: Interview Survey', *British Medical Journal*, **316**, pp. 514-18.

Donnison, D. (1988). 'Defining and Measuring Poverty: a Reply to Stein Ringen', *Journal of Social Policy*, **17**(3), pp. 367-74.

Douglas, A.S., Russell, D. and Allan, T.M. (1990). 'Seasonal, Regional and Secular Variations of Cardiovascular and Cerebrovascular Mortality in New Zealand', *Australian and New Zealand Journal of Medicine*, **20**, pp. 669-76.

Duncan, G.J., Gustafsson, B., Hauser, R., Schmauss, G., Messinger, H., Muffels, R., Nolan, B. and Ray, J.-C. (1993). 'Poverty Dynamics in Eight Countries'. *Journal of Population Economics*, **6**, pp. 215-34.

Edgar, D., Keane, D. and McDonald, P. (eds.) (1989). *Child Poverty*, Allen and Unwin, Sydney.

Ekins, P., Russell, A. and Hargreaves, C. (2001). *Household Energy Efficiency to 2020: the Economic and Environmental Implications of Improving Household Energy Efficiency in the UK*, Report for the Energy Saving Trust, Forum for the Future, London.

Elia, M. (2001). 'Obesity in the Elderly', *Obesity Research*, **9**(4), pp. 244S-48S.

Elola, J., Daponte, A. and Navarro, V. (1995). 'Health Indicators and The Organisation Of Health Care Systems in Western Europe', *American Journal of Public Health*, **85**(10), pp. 1397-01.

Energy Saving Trust (1994). *Recommendations on the Standards of Performance in Energy Efficiency for the Regional Electricity Companies*, Office of Electricity Regulation, London.

Eng, H. and Mercer, J.B. (1998). 'Seasonal Variations in Mortality Caused by Cardiovascular Disease in Norway and Ireland', *Journal of Cardiovascular Risk*, **5**(2), pp. 89-95.

Eurostat (1996). *The European Community Household Panel (ECHP): Survey Methodology and Implementation*, Office for Official Publications of the European Union, Luxembourg.

Eurostat (1999). *Energy Consumption in Households*, Office for Official Publications of the European Union, Luxembourg.

Eurowinter Group (1997). 'Cold Exposure and Winter Mortality from Ischaemic Heart Disease, Cerebrovascular Disease, Respiratory Disease, and All Causes in Warm and Cold Regions of Europe', *Lancet*, **349**, pp. 1341-46.

Evans, J., Hyndman, S., Stewart-Brown, S., Smith, D. and Petersen, S. (2000). 'An Epidemiological Study of the Relative Importance of Damp Housing in Relation to Adult Health', *Journal of Epidemiology and Community Health*, **43**, pp. 677-86.

Eversley, D. and Begg, I. (1984). *Deprivation in the Inner-city – Social Indicators from the 1981 Census*, Economic and Social Research Council, London.

Fahey, T. (ed.). (1999). *Social Housing in Ireland: A Study of Success, Failure and Lessons Learned*, Oak Tree Press, Dublin.

Fajth, G. and Zimakova, T. (1997). 'Family Policies in Eastern Europe: from Socialism to the Market', in G.M. Cornia and M. Danziger (eds.), *Child Poverty in Industrialised Countries*, University Press, Oxford.

Fanger, P.O. (1972). *Thermal Comfort*, McGraw Hill, New York.

Flegg, A.T. (1982). 'Inequality of Income, Illiteracy and Medical Care as Determinants of Infant Mortality in Developing Countries', *Population Studies*, **36**, pp. 441-58.

Forsyth, A., Macintyre, S. and Anderson, A. (1994). 'Diets for Disease? Intra-urban Variation in Reported Food Consumption in Glasgow', *Appetite*, **22**(3), pp. 259-74.

Franz, S.A. and Weaver, E.M. (1994). *Appropriate Treatment of Free Riders in Impact Evaluations: What is a Free Rider Anyway?*, Proceedings of the 1994 ACEEE Summer Study on Energy Efficiency in Buildings.

Gemmell, I., McLoone, P., Boddy, F.A., Dickinson, G.J. and Watt, G.C.M. (2000). 'Seasonal Variation in Mortality in Scotland', *International Journal of Epidemiology*, **29**, pp. 274-79.

Glendinning, C. and Millar, J. (eds.). *Women in Poverty in Britain*, Wheatsheaf Books, London.

Goldberg, E.M., and Morrison, S.L. (1963). 'Schizophrenia and Social Class', *British Journal of Psychiatry*, **109**, pp. 785.

Goode, J., Callender, C. and Lister, R. (1998). *Purse or Wallet? Gender Inequalities and Income Distribution within Families on Benefits,* Policy Studies Institute, London.

Goodwin, J. (2000). 'Cold Stress, Circulatory Illness and the Elderly', in J. Rudge and F. Nicol (eds.), *Cutting the Cost of Cold: Affordable Warmth for Healthier Homes,* E&FN Spon, London.

Gordon, D. (1995). 'Census-based Deprivation Indices: Their Weighting and Validation', *Journal of Epidemiology and Community Health,* **49** (Suppl. 2), pp. S39-S44.

Gordon, D. and Pantazis, C. (1997). *Breadline Britain in the 1990s,* Ashgate, Aldershot.

Gordon, D., Adelman, L., Ashworth, K., Bradshaw, J., Levitas, R., Middleton, S., Pantazis, C., Pastios, D., Payne, S., Townsend, P. and Williams, J. (2000). *Poverty and Social Exclusion in Britain,* Joseph Rowntree Foundation, York.

Gore, C. and Figueireido, J.B. (1996). *Social Exclusion and Anti-Poverty Strategies,* International Institute for Labour Studies, Geneva.

Gravelle, H., Wildman, J. and Sutton, M. (2000). *Income, Income Inequality and Health: What Can We Learn From The Aggregate Data?,* Centre for Health Economics, University of York.

Green, G., Ormandy, D., Brazier, J. and Gilbertson, J. (2000). 'Tolerant Building: the Impact of Energy-Efficiency Measures on Living Conditions and Health Status', in J. Rudge and F. Nicol (eds.), *Cutting the Cost of Cold: Affordable Warmth for Healthier Homes,* E&FN Spon, London.

Green, M.S., Harari, G. and Kristal-Boneh, E. (1994). 'Excess Winter Mortality from Ischaemic Heart Disease and Stroke During Colder and Warmer Years in Israel – An Evaluation and Review of the Role of Environmental Temperature', *European Journal of Public Health,* **4**, pp. 3-11.

Gustafsson, B. (1995). 'Assessing Poverty: Some Reflections on the Literature', *Journal of Population Economics,* **8**, pp. 361-81.

Guy, W.A. (1858). 'On the Annual Fluctuations in the Number of Deaths from Various Diseases, Compared with Like Fluctuations in Crime and in Other Events Within and Beyond the Control of Human Will', *Journal of the Statistical Society,* **21**, pp. 52-86.

Guy, W. A. (1881). 'On Temperature and its Relation to Mortality: an Illustration of the Application of the Numerical Method to the Discovery of Truth', *Journal of the Statistical Society,* **44**, pp. 235-62.

Halleröd, B. (1995a). 'Perceptions of Poverty in Sweden', *Scandinavian Journal of Social Welfare,* **4**(3), pp. 174-89.

Halleröd, B. (1995b). 'The Truly Poor: Indirect and Direct Measurement of Consensual Poverty in Sweden', *Journal of European Social Policy,* **5**(2), pp. 111-29.

Halleröd, B. (1996). 'Deprivation and Poverty: a Comparative Analysis of Sweden and Great Britain', *Acta Sociologica,* **39**, pp. 141-68.

Hartman, R.S. (1988). 'Self-Selection Bias in the Evaluation of Voluntary Energy-Conservation Programs', *Review of Economics and Statistics,* **55**, pp. 448-58.

Harwin, J. and Fajth, G. (1998). 'Child Poverty and Social Exclusion in Post-Communist Societies', *IDS Bulletin,* **29**(1), pp. 66-76.

Hassett, K.A. and Metcalf, G.E. (1992). *Energy Tax Credits and Residential Conservation Investmen,* National Bureau of Economic Research Working Paper #4020.

Haugland, T. (1996). 'Social Benefits of Financial Investment Support in Energy Conservation Policy', *Energy Journal,* **17**(2), pp. 79-102.

Hauser, R. and Fischer, I. (1990). 'Economic Well-Being among Lone-Parent Families', in T. Smeeding, M. O'Higgins and L. Rainwater (eds.), *Poverty, Inequality and Income Distribution in Comparative Perspective,* Harvester Wheatsheaf, London.

Hayes, M.G. (1986). *Urban Decline and Deprivation: Liverpool's Relative Position – A Technical Study,* City Planning Department, Liverpool.

Healy, J.D. (2001). 'Home Sweet Home? Assessing Housing Conditions and Fuel Poverty in Europe', Paper presented at the *31st International Conference on 'Making Cities Livable'*, San Francisco (CA) 22nd-26th October 2001.

Healy, J.D. (2002a). 'Housing Conditions, Energy Efficiency, Affordability and Satisfaction with Housing: a Pan-European Analysis', *Housing Studies*, **18**(3), pp. 409-424.

Healy, J.D. (2002b). 'Poverty, Multiple Deprivation and Social Exclusion in Europe: A Longitudinal Analysis Employing the European Community Household Panel', *mimeo*, University College, Dublin.

Healy, J.D. (2002c). *Lone-Parent Fuel Poverty in Ireland*, One Parent Exchange Network (OPEN), Dublin.

Healy, J.D. (2002d). 'Domestic Energy Efficiency as a Means towards Sustainable Energy Use and Avoiding Non-Compliance', *Proceedings of the 2002 Annual Meeting of the Alliance for Global Sustainability*, San José, Costa Rica.

Healy, J.D. (2003a). 'Fuel Poverty in Europe', *Energy Action*, **89**, pp. 20-21.

Healy, J.D. (2003b). 'Excess Winter Mortality in Europe: Identifying Key Risk Factors', *Journal of Epidemiology & Community Health*, **57**(9), pp. 784-789.

Healy, J.D. (2003c). 'Do Fuel-Poor Households Exhibit Higher Risk Factors Associated with Poor Health?', *International Journal of Health Promotion & Education*, **41**(1), pp. 14-20.

Healy, J.D. (2003d). *Fuel Poverty and Policy in Ireland and the European Union*, Studies in Public Policy #12, The Policy Institute, Trinity College, Dublin.

Healy, J.D. and Clinch, J.P. (2002). 'Fuel Poverty, Thermal Comfort and Occupancy: Results of a National Household Survey in Ireland', *Applied Energy*, **73**, pp. 329-343.

Healy, J.D. and Clinch, J.P. (2004). 'Quantifying the Severity of Fuel Poverty, Its Relationship with Poor Housing and Reasons for Non-Investment in Ireland', *Energy Policy*, **32**: 207-220.

Henwood, M. (1997). *Fuel Poverty, Energy Efficiency & Health*, Eaga Charitable Trust, Cumbria.

Holmes, R. (1998). *Use of Geographical Information Systems in the Identification of Fuel Poverty*, Unpublished M.Sc. thesis, De Montfort University.

Holtermann, S. (1975). 'Areas of Deprivation in Great Britain: an Analysis of 1971 Census Data', *Social Trends*, **6**, pp. 43-8.

Howden-Chapman, P., Isaacs, N., Crane, J. and Chapman, R. (1996). 'Housing and Health: the Relationship between Research and Policy', *International Journal of Environmental Health Research*, **6**, pp. 173-85.

Hutton, S. (1991). 'Measuring Living Standards Using Existing National Data Sets', *Journal of Social Policy*, **20**(2), pp. 237-57.

Hutton, S. (1994). 'Men's and Women's Incomes: Evidence from Survey Data', *Journal of Social Policy*, **23**(1), pp. 21-40.

Ilhan, A., Budak, F., Adanir, M., Komsuoglu, S. (1997). 'Meterological Influences on Cerebrovascular Disease in Trabzon, Turkey', *Journal of Neurological Sciences*, **150** (Suppl. 1), pp. S191.

Illsley, R. and Le Grand, J. (1987). 'The Measurement of Inequality in Health', in A. Williams (ed.), *Health and Economics*, Macmillan, London.

Ingham, A. Maw, J. and Ulph, A. (1991). 'Testing for Barriers to Energy Conservation – an Application of a Vintage Model, *Energy Journal*, **12**(4), pp. 41-64.

Isaacs, N. and Donn, M. (1993). 'Health and Housing – Seasonality in New Zealand Mortality', *Australian Journal of Public Health*, **17**(1), pp. 68-70.

Jaffe, A.B. and Stavins, R.N. (1994a). 'Energy Efficient Investments and Public Policy', *Resource and Energy Economics,* **15**(2), pp. 43-65.

Jaffe, A.B. and Stavins, R.N. (1994b). 'The Energy Paradox and the Diffusion of Conservation Technology', *Resource and Energy Economics*, **16**, pp. 91-122.

Joseph, K. and Sumpton, J. (1979). *Equality*, John Murray, London.

Judge, K., Mulligan, J.-A. and Benzeval, M. (1997). 'Income Inequality and Population Health', *Social Science and Medicine*, **46**, pp. 567-79.

Kangas, O. and Ritakallio, V.M. (1998). 'Different Methods – Different Results? Approaches to Multidimensional Poverty', in J.J. Andreb (ed.), *Empirical Poverty Research in a Comparative Perspective*, Aldershot, Ashgate.

Kawachi, I., Kennedy, B.P. and Wilkinson, R.G. (1999). *The Society and Population Health Reader. Volume 1: Income Inequality and Health*, New Press, New York.

Kearns, R.A. and Smith, C.J. (1998). 'Housing Stressors and Mental Health among Marginalised Urban Populations', *Area*, **25**(3), pp. 267-78.

Keatinge, W.R. (1986). 'Seasonal Mortality among Elderly People with Unrestricted Home Heating', *British Medical Journal*, **293**, pp. 732-3.

Keatinge, W. and Donaldson, G. (2000). 'Cold Weather, Cold Homes and Winter Mortality, in J Rudge and F. Nicol (eds.) (2000), *Cutting the Cost of Cold: Affordable Warmth for Healthier Homes*, E&FN Spon, London.

Kennedy, B.P., Kawachi, I. and Prothrow-Stith, D. (1996). 'Income Distribution and Mortality: Cross-Sectional Ecological Study of the Robin Hood Index in the United States', *British Medical Journal*, **312**, pp. 1004-7.

Khanom, L. (2000). 'Impacts of Fuel Poverty on Health in Tower Hamlets', in J. Rudge and F. Nicol (eds.), *Cutting the Cost of Cold: Affordable Warmth for Healthier Homes*, E&FN Spon, London.

Khaw, K.-T. (1995). 'Temperature and Cardiovascular Mortality', *Lancet*, **345**, pp. 337-8.

Konadu-Agyemang, K. (2001). 'A Survey of Housing Conditions and Characteristics in Accra, an African city', *Habitat International*, **25**(1), pp. 15-34.

Kondo, K., Iwamoto, T. and Hirano, R. (1997). 'Factors Affecting Longevity', *Southeast Asian Journal of Tropical Medicine and Public Health*, **28**, pp. 88-93.

Korsgaard, J. (1983). 'Mite Asthma and Residency: A Case-Control Study on the Impact of Exposure to House-Dust Mites in Dwellings', *American Review of Respiratory Disease*, **128**, pp. 231-35.

Kozma, A. and Stones, M.J. (1983). 'Predictors of Happiness', *Journal of Gerontology*, **38**(5), pp. 626-28.

Kumar, T.K., Gore, A.P. and Sitaramam, V. (1996). 'Some Conceptual and Statistical Issues on the Measurement of Poverty', *Journal of Statistical Planning and Inference*, 49, pp. 53-71.

Kunst, A.E., Looman, C.W.N. and Mackenbach, J.P. (1991). 'The Decline in Winter Excess Mortality in the Netherlands', *International Journal of Epidemiology*, **20**(4), pp. 971-77.

Laake, K. and Sverre, J.M. (1996). 'Winter Excess Mortality: a Comparison Between Norway and England plus Wales, *Age and Ageing*, **25**, pp. 343-48.

Lawlor, J. (1995). *The Costs and Benefits of Government Investments and Subsidies Applied to Energy Conservation in Existing Buildings (Especially Heating and Insulation)*, Economic and Social Research Institute, Dublin.

Layte, R.T., Fahey, T. and Whelan, C.T. (1999). *Income, Deprivation and Well-Being among Older Irish People*, National Council on Ageing and Older People, Report #55, Dublin.

Layte, R., Maître, B., Nolan, B. and Whelan, C.T. (2000). *Persistent and Consistent Poverty in the 1994 and 1995 Waves of the European Community Household Panel Study*, Seminar Paper Presented 26th October 2000, Economic and Social Research Institute, Dublin.

Le Grand, J. and Rabin, M. (1986). 'Trends in British Health Inequality: 1931-83' in A.J. Culyer and B. Jonsson (eds.), *Public and Private Health Services,* Blackwell, Oxford.

Levitas, R. (1999). *The Inclusive Society,* Macmillan, London.

Lewis, J. and Piachaud, D. (1987). 'Women and Poverty in the Twentieth Century', in C. Glendinning and J. Millar (eds.), *Women in Poverty in Britain,* Wheatsheaf Books, London.

Lewis, J. (1992). *Women in Britain Since 1945,* Blackwell, Oxford.

Lewis, P. (1982). *Fuel Poverty Can Be Stopped,* National Right to Fuel Campaign, Bradford.

Lewis-Fanning, E. (1940). *A Comparative Study of the Seasonal Incidence of Mortality in England and Wales in the United States of America,* HM Stationary Office (Medical Research Council Special Series #239), London.

Lobmayer, P. and Wilkinson, R. (2000). 'Income, Inequality and Mortality in 14 Developed Countries, *Sociology of Health & Illness,* **22**(4), pp. 401-14.

Long, J.E. (1993). 'An Economic Analysis of Residential Expenditure', *Energy Economics,* **15**, pp. 232-8.

Mack, J. and Lansley, S. (1985). *Poor Britain,* Allen and Unwin, London.

Mannino, D.M., Ford, E., Giovino, G.A. and Thun, M. (2001). 'Lung Mortality Rates in Birth Cohorts in the United States from 1960 to 1994', *Lung Cancer,* **31**(2), pp. 91-99.

Mant, D. C., Muir Gray, J. A. (Eds.). (1986). *Building Regulations and Health,* Building Research Establishment, Watford.

Marsh, A., Gordon, D., Pantazis, C. and Heslop, P. (1999). *Home Sweet Home? The Impact of Poor Housing on Health,* Policy Press, Bristol.

McGregor, P.P.L. and Borooah, V.K. (1992). 'Is Low Spending or Low Income a Better Indicator of Whether a Household is Poor: Some Results from the 1985 Family Expenditure Survey', *Journal of Social Policy,* **21**(1), pp. 53-69.

McKee, M., Sanderson, C., Chenet, L., Vassin, S. and Shkolnikov, V. (1998). 'Seasonal Variation in Mortality in Moscow', *Journal of Public Health Medicine,* **20**(3), pp. 268-74.

McMahon, B. (1993). 'Time for a Change of Direction: Effects of Poverty on Ill Health and Service Provision', *Professional Nurse,* **8**(9), pp. 610-13.

McSharry, B. (1993). *Energy Conservation in the Domestic Sector in the Republic of Ireland,* Unpublished M.Sc. thesis, University of Dublin.

Melia, R.J.W., Florey, C.V., Morris, R.W. 'Childhood Respiratory Illness and the Home Environment: Association between Respiratory Illness and Nitrogen Oxides, Temperature and Relative Humidity', *International Journal of Epidemiology,* **11**, pp. 164-9.

Mellor, J.M. and Milyo, J. (2001). 'Re-examining the Evidence of an Ecological Association between Income Inequality and Health', *Journal of Health Politics, Policy and Law,* **26**(3), pp. 487-522.

Micklewright, J. and Stewart, K. (1999). 'Is Child Welfare Converging in the EU?', *Innocenti Occasional Papers,* Economic and Social Policy Series #69, UNICEF, Florence.

Millar, A. (1980). *A Study of Multiply Deprived Households in Scotland,* Central Research Unit Papers, Scottish Office, Edinburgh.

Millar, J. (1990). *Lone-Parent Families in Ireland: Report to the OECD,* University of Bath Centre for the Analysis of Social Policy, Bath.

Milne, G. and Boardman, B. (2000). 'Making Cold Homes Warmer: the Effect of Energy Efficiency Improvements in Low-Income Houses', *Energy Policy,* **28**, pp. 411-24.

Mitchell, D. (1991). *Income Transfers in Ten Welfare States,* Ashgate, Newcastle.

Muffels, R. and de Vries, A. (1989). *Poverty in the Netherlands: First Report of an International Comparative Study,* VUGA, Tilburg.

Muffels, R. and Vreins, M. (1991). 'The Elaboration of a Deprivation Scale and the Definition of a Subjective Deprivation Poverty Line', Paper Presented at the *Annual Meeting of the European Society for Population Economics,* 6th-8th June, Pisa.

Muffels, R., Berghman, J. and Dirven, H. (1992). 'A Multi-Method Approach to Monitor the Evolution of Poverty, *Journal of European Social Policy,* **2**(3), pp. 193-213.

Muller A. (2002). Education, Income Inequality and Mortality: a Multiple Regression Analysis, *British Medical Journal,* **324**, pp. 23-25.

National Economic and Social Forum (NESF) (2000). *The National Anti-Poverty Strategy: Forum: Opinion No. 8,* Government Publications Office, Dublin.

Neighbourhood Energy Action (NEA) (1997). *Fuel Poverty in Northern Ireland,* NEA, Belfast.

Nolan, B. and Whelan, C.T. (1996a). 'Measuring Poverty Using Income and Deprivation Indicators: Alternative Approaches', *Journal of European Social Policy,* **6**(3), pp. 225-40.

Nolan, B. and Whelan, C.T. (1996b). *Resources, Deprivation and Poverty,* Clarendon Press, Oxford.

Nolan, B., Whelan, C.T. and Williams, J. (1998). *Where are Poor Households? The Spatial Distribution of Poverty and Deprivation in Ireland.* Dublin: Oak Tree Press.

Nolan, B. and Watson, D. (1999). *Women and Poverty in Ireland.* Dublin: Combat Poverty Agency.

Nolan, B. (2000). *Child Poverty in Ireland,* Oak Tree Press, Dublin.

Nolan, B. and Russell, H. (2001). 'Non-Cash Benefits and Poverty in Ireland, *Policy Research Series #39,* Economic and Social Research Institute, Oxford.

Oranga, H.M. (1997). 'Ageing and Poverty in Rural Kenya: Community Perception', *East African Medical Journal,* **74**(10), pp. 611-13.

Packer, C.N., Stewart-Brown, S. and Fowle, S. (1994). 'Damp Housing and Adult Health: Results from a Lifestyle Survey in Worcester, England', *Journal of Epidemiology & Community Health,* **48**, pp. 555-9.

Pan, W.-H., Li, L.-A. and Tsai, M.-J. (1995). 'Temperature Extremes and Mortality from Coronary Heart Disease and Cerebral Infarction in Elderly Chinese', *Lancet,* **345**, pp. 353-5.

Paukert, F. (1973). 'Income Distribution at Different Levels of Development: a Survey of Evidence', *International Labour Review,* **108**, pp. 97-125.

Peters, J. and Stevenson, M. (2000). 'Modelling the Health Cost of Cold, Damp Housing, in J. Rudge and F. Nicol (eds.), *Cutting the Cost of Cold: Affordable Warmth for Healthier Homes,* E&FN Spon, London.

Piachaud, D. (1987). 'Problems in the Definition and Measurement of Poverty', *Journal of Social Policy,* **16**(2), pp. 147-64.

Platts-Mills, T.A.E. and Chapman, M.D. (1987). 'Dust Mites: Immunology, Allergic Disease and Environment Control', *Journal of Allergy and Clinical Immunology,* **80**, pp. 755-75.

Poikolainen, K. and Eskola, J. (1988). 'Health Services Resources and their Relation to Mortality from Causes Amenable to Health Care Intervention: a Cross-National Study, *International Journal of Epidemiology,* **17**(1), pp. 86-9.

Ponka, A. and Virtanen, M. (1996). 'Asthma and Ambient Air Pollution in Helsinki', *Journal of Epidemiology and Community Health,* **50** (Suppl. 1), pp. S59-S62.

Preston, S.H. (1975). 'The Changing Relation between Mortality and Level of Economic Development', *Population Studies,* **29**, pp. 231-48.

Pringle, D.G., Walsh, J. and Hennessy, M. (Eds.) (1999). *Poor People, Poor Places: A Geography of Poverty and Deprivation in Ireland*, Oak Tree Press, Dublin.

Quigley, J. M. (1991). 'Residential Energy Conservation: Standards, Subsidies and Public Programs', in R. J. Gilbert (ed.) *Developments in Energy Regulation*, University of California Press, Berkeley.

Quinn, P. (Ed.) (1995). *Energy Conservation and Job Creation in the Domestic Sector*, Energy Enterprises, Dublin.

Rakodi, C. and Withers, P. (1995). 'Housing Aspirations and Affordability in Harare and Gweru: a Contribution to Housing Policy Formulation in Zimbabwe', *Cities*, **12**(3), pp. 185-201.

Ramprakash, D. (1994). 'Poverty in the Countries of the European Union: a Synthesis of Eurostat's Statistical Research on Poverty', *Journal of European Social Policy*, **2**, pp. 117-28.

Raw, G. J. and Hamilton, R. M. (Eds.). (1995). *Building Regulations and Health*, Building Research Establishment, Watford.

Ringen, S. (1988). 'Direct and Indirect Measures of Poverty', *Journal of Social Policy*, **17**(3), pp. 351-65.

Rodgers, G.B. (1979). 'Income and Inequality as Determinants of Mortality: an International Cross-Section Analysis', *Population Studies*, **33**, pp. 343-51.

Rogot, E. and Padgett, S.J. (1976). 'Associations of Coronary and Stroke Mortality with Temperature and Snowfall in Selected Areas of the United States, 1962-66, *American Journal of Epidemiology*, **103**, pp. 565-75.

Room, G. (ed.) (1995). *Beyond the Threshold: the Measurement and Analysis of Social Exclusion*, Policy Press, Bristol.

Rose, G. (1966). 'Cold Weather and Ischaemic Heart Disease', *British Journal of Preventative Social Medicine*, **20**, pp. 97-100.

Rose, A. and Lin, S.-M. (1995). 'Regrets or No Regrets – That Is the Question: Is Conservation a Costless CO_2 Mitigation Strategy?, *Energy Journal*, **16**(3), pp. 67-87.

Rowles, G.D. and Johansson, H.K. (1993). 'Persistent Elderly Poverty in Rural Appalachia', *Journal of Applied Gerontology*, **12**(3), pp. 349-67.

Rowntree, B.S. (1901). *Poverty: A Study of Town Life*, Macmillan & Co, London.

Rowntree, B.S. (1941). *Poverty and Progress: A Second Social Survey of York*, Longman's, Green & Co, London.

Rowntree, B.S. and Lavers, G.R. (1951). *Poverty and the Welfare State: A Third Social Survey of York Dealing Only with Economic Questions*, Longman's, Green & Co, London.

Rudge, J. (2000). 'Winter Morbidity and Fuel Poverty: Mapping the Connection', in J. Rudge and F. Nicol (eds.), *Cutting the Cost of Cold: Affordable Warmth for Healthier Homes*, E&FN Spon, London.

Rudge, J. and Nicol, F. (2000). *Cutting the Cost of Cold: Affordable Warmth for Healthier Homes*, E & FN Spon, London.

Saez, M., Sunyer, J., Castellsague, J. Murillo, C and Anto, J.M. (1995). 'Relationship between Weather Temperature and Mortality: a Time-Series Analysis Approach in Barcelona', *International Journal of Epidemiology*, **24**, pp. 576-82.

Salvage, A.V. (1992). *Energy Wise? Elderly People and Domestic Energy Efficiency*, Age Concern, London.

Salvaggio, J. and Aukrust, L. (1981). 'Mold-Induced Asthma', *Journal of Allergy and Clinical Immunology*, **68**(5), pp. 327-46.

Sckumatz, L. A. (1996). *Recognising All Programme Benefits: Estimates of Non-Energy Benefits from the Customer Perspective*, Skumatz Economic Research Associates Inc., Washington.

Scott, S. (1993). 'Market Failure and Energy Conservation in the Home', Paper Presented at the Conference, *Irish Energy Policy in a European Context*, 8th March 1993, Economic and Social Research Institute, Dublin.

Scott, S. (1995). 'The Assumption of Rational Behaviour: Energy Efficiency Investments in the Home', Paper Presented at the *Annual Conference of the Irish Economics Association*, Co. Cavan, Ireland.

Scott, S. (1997). 'Household Energy Efficiency in Ireland; a Replication Study of Ownership of Energy-Saving Items, *Energy Economics*, **19**, pp. 197-208.

Sefton, T. (2002). 'Cost-Effectiveness of Policies to Tackle Fuel Poverty: the Case of HEES', *Fiscal Studies*, **23**(3), pp. 369-399.

Sen, A.K. (1973). *On Income Inequality*, Clarendon Press, Oxford.

Sen, A.K. (1983). 'Poverty Relatively Speaking', *Oxford Economic Papers,* **35**, pp. 153-69.

Seto, T., Mittlemann, M., Davis, R., Taira, D. and Kawachi, I. 'Seasonal Variation in Coronary Artery Disease in Hawaii: Observational Study', *British Medical Journal*, **316**, pp. 1,946-47.

Shaw, S. and Peacock, J. (1999). 'Deprivation and Excess Winter Mortality', *Journal of Epidemiology and Community Health*, **53**, pp. 499-502.

Shibuya, K., Hashimoto, H. and Yano, E. (2002). 'Individual Income, Income Distribution and Self-Rated Health in Japan: Cross-Sectional Analysis of Nationally Representative Sample, *British Medical Journal*, **324**, pp. 16.

Silver, H. (1994). 'Social Exclusion and Social Solidarity: Three Paradigms', *International Labour Review*, **133**(6), pp. 1-30.

Smeeding, T., Rainwater, L., Rein, M., Hauser, R. and Schaber, G. (1990). 'Income Poverty in Seven Countries: Initial Estimates from the LIS Database', in Smeeding, T., O'Higgins, M., Rainwater, L. (eds.) *Poverty, Inequality and Income Distribution in Comparative Perspective*, Harvester Wheatsheaf, Hemel Hempstead.

Smeeding, T. and Torrey, B.B. (1988). 'Poor Children in Rich Countries', *Science*, **242**, pp. 873-77.

Smith, S. (1992). 'The Distributional Consequences of Taxes on Energy and the Carbon Content of Fuels, in "The Economics Of Limiting CO_2 Emissions"', *European Economy*, Special Edition (**1**), pp. 241-68.

Strachan, D. and Elton, R. 'Relationship between Respiratory Morbidity in Children and the Home Environment', *Family Practitioner,* **3**, pp. 137-42.

Szalai, J. (1992). 'Social Policy and Child Poverty: Hungary Since 1945', *Innocenti Occasional Papers Economic and Policy Series* #32, International Child Development Centre, UNICEF, Florence.

Szulc, A. (1995). 'Measurement of Poverty: Poland in the 1980s', *Review of Income and Wealth*, **41**(2), pp. 191-205.

Tadesse, S. (2000) (United Nations HQ). *Personal communication*.

Thompson, H., Petticrew, M. and Morrison, D. (2001). 'Health Effects of Housing Improvement: Systematic Review of Intervention Studies', *British Medical Journal*, **323**, pp. 187-90.

Thompson, P.B. (1997). 'Evaluating Energy Efficiency Investments: Accounting for Risk in the Discounting Process', *Energy Policy*, **25**(12), pp. 989-96.

Tietenberg, T. (1997). *Information Strategies for Pollution Control,* Paper Presented at the 8th Annual European Association of Environmental and Resource Economists, Tilburg University.

Touloumi, G., Pocock, S.J., Katsouyanni, K. and Trichopoulos, D. (1997). 'Short-Term Effects of Air Pollution on Daily Mortality in Athens: a Time-Series Analysis', *International Journal of Epidemiology*, **23**(5), pp. 957-67.

Townsend, P. (1962). 'The Meaning of Poverty', *British Journal of Sociology*, **8**, pp. 347-68.
Townsend, P. (1970). *The Concept of Poverty*, Heinmann, London.
Townsend, P. (1979). *Poverty in the United Kingdom*, Penguin Books, London.
Townsend, P. (1987). Deprivation, *Journal of Social Policy*, **16**, pp. 125-46.
Trilling, L. (1950). *The Liberal Imagination: Essays on Literature and Society*, Secker & Warburg, London.
UNICEF (1997). *Children at risk in Central and Eastern Europe: perils and promises*, Regional Monitoring Report #4, International Child Development Centre, UNICEF, Florence.
UNICEF (2000). *A League Table of Child Poverty in Rich Nations*, Innocenti Report Card Issue #1, UNICEF Innocenti Research Centre, Florence.
Van den Bosch, K. (1998). 'Perceptions of the Minimum Living Standard of Living in Belgium: Is There a Consensus?', in Andreb, H.J. (ed.) *Empirical Poverty Research in a Comparative Perspective*, Ashgate, Aldershot.
Vaughan, D.R. (1993). 'Exploring the Use of the Public's Views to Set Income Poverty Thresholds and Adjust Them Over Time', *Social Security Bulletin*, **56**(2), pp. 22-46.
Veit-Wilson, J.H. (1987). 'Consensual Approaches to Poverty Lines and Social Security', *Journal of Social Policy*, **16**(2), pp. 183-211.
Visscher, T.L., Seidell, J.C., Molarius, A., van der Kuip, D., Hofman, A., Witteman, J.C. (2001). 'A Comparison of Body Mass Index, Waist-Hip Ratio and Waist Circumference as Predictors of All-Cause Mortality among the Elderly: the Rotterdam Study', *International Journal of Obesity and Related Metabolic Disorders*, **25**(11), pp. 1730-35.
Vogel, J. (1997). *Living conditions and inequality in the European Union in 1997*, Eurostat Working Papers, Population and Social Conditions E/1997-3, Eurostat, Luxembourg.
Waldman, R.J. (1992). 'Income Distribution and Infant Mortality', *Quarterly Journal of Economics*, **107**, pp. 1,283-1,302.
Waldman, D.M. and Ozog, M.T. (1996). 'Natural and Incentive-Induced Conservation in Voluntary Energy Management Programs', *Southern Economic Journal*, **62**(4), pp. 1,054-71.
Walker, R. (1987). 'Consensual Approaches to the Definition of Poverty: Towards an Alternative Methodology', *Journal of Social Policy*, **16**(2), pp. 213-26.
Walsh, M.J. (1989). 'Energy Tax Credits and Housing Improvements', *Energy Economics*, **11**(4), pp. 275-85.
Waitzkin, H. (1998). 'Is Our Work Dangerous? – Should It Be?' *Journal of Health and Social Behaviour*, **39**(1), pp. 7-17.
Wang, J., Jamison, D.T. Bos, E. and Yu, M.T. (1997). 'Poverty and Mortality among the Elderly: Measurement of Performance in 33 countries 1960-92', *Tropical Medicine and International Health*, **2**(10), pp. 1,001-10.
Watt, G.C. (1994). 'Health Implications of Putting Value Added Tax on Fuel: Time to Combat Fuel Poverty', *British Medical Journal*, **309**, pp. 1029-30.
Weber, G. (1990). *Earnings-Related Borrowing Restrictions: Empirical Evidence from a Pseudo Panel for the UK*, Department of Economics Discussion Paper 90-17, University College London.
Wedgewood, J. (1929). *The Economics of Inheritance*, Penguin, Harmondsworth.
West, R.R., Lloyd, S. and Roberts, C.J. (1973). 'Mortality from Ischaemic Heart Disease: Association with Weather', *British Journal of Preventative Social Medicine*, **27**, pp. 36-40.
Whelan, C.T. (1996). 'Marginalization, Deprivation and Fatalism in the Republic of Ireland: Class and Underclass Perspectives', *European Sociological Review*, **12**(1), pp. 33-51.

Whyley, C. and Callender, C. (1997). *Fuel Poverty in Europe: Evidence from the European Household Panel Survey*, Policy Studies Institute, London.

Wicks, M. (1978). *Old and Cold: Hypothermia and Social Policy*, Heinemann, London.

Wildman J. (2001). 'The Impact of Income Inequality on Individual and Societal Health: Absolute Income, Relative Income and Statistical Artefacts', *Health Economics*, **10**(4), pp. 357-62.

Wilkinson, P., Landon, M. and Stevenson, S. (2000). 'Housing and Winter Death: Epidemiological Evidence', in J. Rudge and F. Nicol (eds.) (2000), *Cutting the Cost of Cold: Affordable Warmth for Healthier Home,* E&FN Spon, London.

Wilkinson, R.G. (1989). 'Class Mortality Differentials, Income Distribution and Trends in Poverty 1921-81', *Journal of Social Policy,* **18**, pp. 307-35.

Wilkinson, R.G. (1992). 'Income Distribution and Life Expectancy', *British Medical Journal*, **304**, pp. 165-68.

Wilkinson, R.G. (1994). 'The Epidemiological Transition: from Material Scarcity to Social Disadvantage?', *Daedalus*, **123**, pp. 61-77.

Wilkinson, R.G. (1996). *Unhealthy Societies: The Afflictions of Inequality.* Routledge, London.

Wilkinson, R.G. (1997). 'Income Inequality Summarises the Health Burden of Individual Relative Deprivation', *British Medical Journal,* **313**, pp. 1,727-28.

Wilkinson, R.G. (2000). *Mind the Gap: Hierarchies, Health and Human Evolution,* Weidenfeld and Nicolson, London.

Williams, R. and Ross, M. (1980). 'Drilling for Oil and Gas in our Houses', *Technology Review*, **82**(5), pp. 24-36.

Williamson, I.J., Martin, C.J., McGill, C., Monie, R.D.H. and Fennerty, A.D. (1997). 'Damp Housing and Asthma: a Case-Control Study', *Thorax,* **52**, pp. 229-34.

Wilmshurst, P. (1994). 'Temperature and Cardiovascular Mortality', *British Medical Journal*, **309**, pp. 309-10.

Wolfson, M., Kaplan, G., Lynch, N.R. and Baukland, E. (1999). 'Relation between Income Inequality and Mortality: Empirical Demonstration', *British Medical Journal*, **319**, pp. 953-5.

World Health Organisation (WHO) (2002). http://www.who.int/dsa/

Index

Austria 2, 9-10, 12, 19, 21, 24, 30, 33-4, 36, 40, 44, 103, 144, 181, 190-1, 198, 206

Belgium 2, 9, 19, 21, 23-4, 30, 43, 47-9, 63, 101, 142-4, 163, 181, 187, 190, 192-3, 199, 206
Boardman, B. 4, 22, 32, 33, 35, 58-9, 66-7, 87, 91, 131, 134-5, 169

Collins, K.J. 7, 34, 79, 89, 90, 92, 121, 129, 131, 134-6
Cooling 61-2
cost-benefit analysis 84, 166, 168

Denmark 2, 9, 11, 13-14, 17-19, 22, 24-5, 31, 34, 40-1, 45, 141, 144, 147, 162-3, 177, 181, 190, 192

energy
   consumption 3-4, 8, 22-3, 33-4, 57, 77, 176, 184, 186
   efficiency 7, 15, 21-2, 29-30, 34, 66, 82-4, 89, 96, 106, 114, 127-8, 158, 161-2, 165-90
   Home Energy Efficiency Scheme (HEES) 161, 186-7
European Community Household Panel (ECHP) 2, 6, 8-10, 33-4, 36-40, 88, 143-4, 152, 185-6
Eurostat 8, 9, 17, 21, 34, 36-9, 144
excess winter mortality
   climate 145-7
   definition 142
   EU 145
   Healthcare 149-50
   housing standards 156-7
   lifestyle risk factors 150-52
   macroeconomic factors 147-8
   risk factors 145-57
   socio-economic variables 152-5

Finland 2, 9, 10, 12-13, 21, 27, 31, 34, 36, 45, 141, 144-5, 156, 162-3, 179, 181, 185, 190, 192

France 2, 9, 10, 12-15, 18-20, 24, 26-7, 30, 31, 33, 38, 40-2, 45, 47-51, 53, 56-8, 61, 63, 102-3, 142-4, 149, 163-5, 178, 182, 187, 190-3, 198, 206
fuel allowance 32, 78-9, 86, 164, 174, 178, 183, 185, 188
fuel poverty
   chronic, intermittent 67-86
   consensual approach 35-6, 39, 58-9
   definition 3-4, 32-3
   EU 33-4, 36-58
   measurement 35-36
   objective indicators 43-6
   occupancy 137-8
   severity 67-8
   subjective indicators 40-3

Germany 2, 9, 10, 12-18, 24, 27-8, 30-1, 37-8, 40, 61, 92, 97, 141-5, 147, 149, 162, 177, 182, 184, 190-2
Greece 2, 3, 9, 12, 15-16, 18-34, 37-8, 40-3, 46-50, 54-8, 63, 93-9, 102-4, 163, 165, 182-92, 198, 203, 206

health
   chronic outcomes 113-4
   cold stress 113, 124, 129-39, 140, 146, 158, 160
   diet 124-5
   dynamic effects 108
   emotional 110-11
   fuel poverty 90-92, 106-18, 152-5
   housing affordability 99-101
   housing conditions 90-9
   housing satisfaction 101-3
   lifestyle risk factors 150-2
   mental 109-10
   morbidity costs 164, 166, 180
   objective outcomes 111-12
   physical activity 125
   physical outcomes 108-9
   quality of life 115-6
   self-perceived 111, 114
   shivering 121-23

worries 114-5
Healy, J.D. 3-5, 7, 29-30, 33-4, 44-5, 58, 64-6, 77, 82-4, 88-9, 98, 105-6, 116, 119, 128-9, 131, 138-41, 144, 146, 156, 162, 164-8, 180, 183
housing
  affordability 8, 23-5
  age 71-2
  condensation 79-81
  conditions 7-8, 12-23, 66, 75, 79-81, 88
  damp 8, 15, 19-22, 29, 36, 39, 43-4, 47, 61, 77-80, 90, 93-4, 112
  deprivation 7-31
  overcrowding 8, 17-9, 29, 90, 98-9, 103, 105, 163
  rotten windows 20-1, 45-6, 94-5
  satisfaction 8, 23, 27-9, 99-103
  tenure 10, 12-3, 50, 57-61, 73, 77-8, 82
  type 11-2

Ireland 1-14, 17, 19, 21, 23-7, 29-34, 36-8, 40-1, 44, 46-9, 51, 53-5, 57-8, 63-86, 88-9, 93-7, 104-139, 144-7, 149, 152, 156, 158-9, 161-76, 180, 183-99, 201, 203-4, 206-7
Italy 2-3, 9-10, 12-14, 16-19, 21, 23-6, 28-9, 31, 34, 37-8, 40-1, 43, 47, 48-9, 54, 57-8, 61, 63, 93-6, 102, 104, 116, 141-5, 163, 184, 190, 192-3, 198-9, 206

market failure
  income constraints 83-4, 167
  information asymmetry 167
  information gap 83, 167-8, 170-1
  property rights 84, 169
  transactions' costs 84-6, 168

regression analysis
  multivariate 60-2, 81-2
  Poisson 158-9
  Probit 60-2, 81-2, 222-34

self-reported data
  fuel poverty 39-46
  health 89-92, 98-102, 105-8, 116, 133, 135, 138-9, 163, 192, 199, 201-2, 205
  housing conditions 15, 18, 19, 23, 27, 29, 31
sensitivity analysis 33, 46-9, 63
social exclusion 9, 36
socio-demographic analysis 50-9, 68-73
socio-economic analysis 50-9, 73-9, 81, 83, 139, 152-8
subsidy 78, 86, 173, 176-7, 185, 190

taxation 166-7, 170, 177-89
thermal comfort
  and cold stress/strain 136-7
  definition 131-2
  and fuel poverty 132-3
  and temperature 134-5
Townsend, P. 33, 35, 87, 127, 140-1

Wilkinson, R.G. 141
World Health Organisation (WHO) 131, 134, 139

Made in the USA
Las Vegas, NV
12 October 2022